■ Simulation and Similarity

OXFORD STUDIES IN PHILOSOPHY OF SCIENCE

General Editor: Paul Humphreys, University of Virginia

Advisory Board
Anouk Barberousse (European Editor)
Robert Batterman
Jeremy Butterfield
Peter Galison
Philip Kitcher
Margaret Morrison
James Woodward

The Book of Evidence
Peter Achinstein

Science, Truth, and Democracy
Philip Kitcher

Inconsistency, Asymmetry, and Non-Locality
A Philosophical Investigation of Classical Electrodynamics
Mathias Frisch

The Devil in the Details: Asymptotic Reasoning in Explanation, Reduction, and Emergence
Robert W. Batterman

Science and Partial Truth: A Unitary Approach to Models and Scientific Reasoning
Newton C.A. da Costa and Steven French

Inventing Temperature: Measurement and Scientific Progress
Hasok Chang

The Reign of Relativity: Philosophy in Physics 1915 – 1925
Thomas Ryckman

Making Things Happen
James Woodward

Mathematics and Scientific Representation
Christopher Pincock

Simulation and Similarity
Michael Weisberg

Simulation and Similarity

Using Models to Understand the World

Michael Weisberg

UNIVERSITY PRESS

Oxford University Press is a department of the University of Oxford.
It furthers the University's objective of excellence in research, scholarship,
and education by publishing worldwide.

Oxford   New York
Auckland   Cape Town   Dar es Salaam   Hong Kong   Karachi
Kuala Lumpur   Madrid   Melbourne   Mexico City   Nairobi
New Delhi   Shanghai   Taipei   Toronto

With offices in
Argentina   Austria   Brazil   Chile   Czech Republic   France   Greece
Guatemala   Hungary   Italy   Japan   Poland   Portugal   Singapore
South Korea   Switzerland   Thailand   Turkey   Ukraine   Vietnam

Oxford is a registered trade mark of Oxford University Press
in the UK and certain other countries.

Published in the United States of America by
Oxford University Press
198 Madison Avenue, New York, NY 10016

© Michael Weisberg 2013

First issued as an Oxford University Press paperback, 2015.

All rights reserved. No part of this publication may be reproduced, stored in a
retrieval system, or transmitted, in any form or by any means, without the prior
permission in writing of Oxford University Press, or as expressly permitted by law,
by license, or under terms agreed with the appropriate reproduction rights organization.
Inquiries concerning reproduction outside the scope of the above should be sent to the
Rights Department, Oxford University Press, at the address above.

You must not circulate this work in any other form
and you must impose this same condition on any acquirer.

Library of Congress Cataloging-in-Publication Data
Weisberg, Michael, 1976–
  Simulation and similarity : using models to understand the world / Michael Weisberg.
    p. cm. – (Oxford studies in philosophy of science)
  Includes bibliographical references (p.).
  ISBN 978-0-19-993366-2 (hardback : alk. paper); 978-0-19-026512-0 (paperback : alk. paper)
  ISBN 978-0-19-993367-9 (e-book)
  1. Science–Mathematical models–Philosophy. I. Title.
  Q175.32.M38W45   2013
  003–dc23   2012023391

In time, those Unconscionable Maps no longer satisfied, and the Cartographers Guilds struck a Map of the Empire whose size was that of the Empire, and which coincided point for point with it ... In the Deserts of the West, still today, there are Tattered Ruins of that Map, inhabited by Animals and Beggars;

On Exactitude in Science
Jorge Luis Borges

For Deena

CONTENTS

List of Figures		xiii
List of Tables		xv
Preface		xvii

1 Introduction — 1
  1.1 Two Aquatic Puzzles — 1
  1.2 Models of Modeling — 4

2 Three Kinds of Models — 7
  2.1 Concrete Model: The San Francisco Bay–Delta Model — 7
  2.2 Mathematical Model: Lotka–Volterra Model — 10
  2.3 Computational Model: Schelling's Segregation Model — 13
  2.4 Common Features of these Models — 14
  2.5 Only Three Types of Models? — 15
  2.6 Fewer than Three Types of Model? — 19

3 The Anatomy of Models — 24
  3.1 Structure — 24
    3.1.1 Concrete Structures — 24
    3.1.2 Mathematical Structures — 25
    3.1.3 Computational Structures — 29
  3.2 Model Descriptions — 31
  3.3 Construal — 39
  3.4 Representational Capacity of Structures — 42

4 Fictions and Folk Ontology — 46
  4.1 Against Maths: Individuation, Causes, and Face-Value Practice — 46
  4.2 A Simple Fictions Account — 49
  4.3 Enriching the Simple Account — 51
    4.3.1 Waltonian Fictionalism — 53
    4.3.2 Fictions Without Models — 55
  4.4 Why I Am Not a Fictionalist — 56
    4.4.1 Variation — 56
    4.4.2 Representational Capacity of Different Models — 61
    4.4.3 Making Sense of Modeling — 63
    4.4.4 Variation in Practice — 64

	4.5	Folk Ontology	67
	4.6	Maths, Interpretation, and Folk Ontology	70
5	Target-Directed Modeling		74
	5.1	Model Development	75
	5.2	Analysis of the Model	79
		5.2.1 Complete Analysis	79
		5.2.2 Goal-Directed Analysis	83
	5.3	Model–Target Comparison	90
		5.3.1 Phenomena and Target Systems	90
		5.3.2 Establishing the Fit between Model and Target	93
		5.3.3 Representations of Targets	95
6	Idealization		98
	6.1	Three Kinds of Idealization	98
		6.1.1 Galilean Idealization	99
		6.1.2 Minimalist Idealization	100
		6.1.3 Multiple-Models Idealization	103
	6.2	Representational Ideals and Fidelity Criteria	105
		6.2.1 COMPLETENESS	105
		6.2.2 SIMPLICITY	107
		6.2.3 1-CAUSAL	107
		6.2.4 MAXOUT	109
		6.2.5 P-GENERAL	109
	6.3	Idealization and Representational Ideals	110
	6.4	Idealization and Target-Directed Modeling	112
7	Modeling Without a Specific Target		114
	7.1	Generalized Modeling	114
		7.1.1 How-Possibly Explanations	118
		7.1.2 Minimal Models and First-Order Causal Structures	119
	7.2	Hypothetical Modeling	121
		7.2.1 Contingent Nonexistence: xDNA	122
		7.2.2 Impossible Targets: Infinite Population Growth and Perpetual Motion	124
	7.3	Targetless Modeling	129
	7.4	A Moving Target: The Case of Three-sex Biology	131
8	An Account of Similarity		135
	8.1	Desiderata for Model–World Relations	135
	8.2	Model-Theoretic Accounts	137
	8.3	Similarity	142
	8.4	Tversky's Contrast Account	143

	8.5	Attributes and Mechanisms	145
	8.6	Feature Sets, Construals, and Target Systems	148
	8.7	Modeling Goals and Weighting Parameters	150
	8.8	Weighting Function and Background Theory	152
	8.9	Satisfying the Desiderata	154
9	Robustness Analysis and Idealization		156
	9.1	Levins and Wimsatt on Robustness	156
	9.2	Finding Robust Theorems	158
	9.3	Three Kinds of Robustness	159
		9.3.1 Parameter Robustness	160
		9.3.2 Structural Robustness	161
		9.3.3 Representational Robustness	162
	9.4	Robustness and Confirmation	167
10	Conclusion: The Practice of Modeling		171

References   176
Index   186

LIST OF FIGURES

1.1	A schematic of the Reber plan.	2
2.1	Lotka–Volterra model's oscillation.	12
2.2	An example of Schelling's model of segregation on a 51 × 51 grid with 2000 agents. Each agent prefers 30% of its Moore neighbors to be the same shape and color. The initial distribution of agents was random, and the model equilibrated after fourteen time steps.	14
3.1	This representation of the phase space for the Lotka–Volterra model.	28
3.2	Technical drawing of the San Francisco Bay model, showing the model's scale (1:10,000) and orientation. The portion of the model representing the Suisun Bay and San Joaquin Delta was rotated 43 degrees so that it could fit in the warehouse.	32
3.3	A segment of the San Francisco Bay model, showing its representation of the Golden Gate Bridge.	33
3.4	Model description for a family of Lotka–Volterra models with the assignment made explicit.	40
5.1	Left: a Tasmanian devil in captivity; photograph taken by the author. Right: a depiction of how abstracting away from a single phenomenon can yield many different kinds of targets.	91
5.2	A set of observations for Canadian lynx population sizes fit to the Lotka–Volterra model.	94
5.3	The relationship between phenomenon, target, and model for a concrete model.	96
5.4	The relationship between phenomenon, target, mathematical representation of target, and model for a mathematical model.	96
7.1	Left: Patterns of racial segregation in Philadelphia from the 2010 census. Darker areas correspond to higher percentages of African Americans. Right: The output of a typical run of Schelling's segregation model using a virtual city that is 51 × 51 and two types of agents (gray and white). All agents prefer to have at least 30% same-colored neighbors.	119
7.2	Structures of the size-expanded nucleosides benzoadenine and benzothymine.	123
7.3	Feynman's ratchet illustrated by Zhanchun Tu. Courtesy of Professor Tu.	126

7.4	A glider in the game of life. This structure persists and will "move" across the game board.	130
8.1	These two finite graphs are isomorphic to one another.	138
8.2	Left: The equilibrium state of a typical run of Schelling's segregation model. Right: A linear transformation of this state, which is isomorphic to the original.	140
9.1	Output of Weisberg and Reisman's density-dependent individual-based predation model with biocide perturbation.	166

LIST OF TABLES

3.1 Scale relationships for the San Francisco Bay model   36

PREFACE

My interest in models, modeling, and idealization began when I first learned about the strictly incompatible and highly idealized models chemists use to describe bonding behavior. The practice seemed puzzling; why didn't chemists just use the best model? This book is an attempt to synthesize fifteen years' worth of thinking about this question and many others that have arisen along the way. It draws on articles previously published in *The British Journal for Philosophy of Science, Philosophy of Science,* and *The Journal of Philosophy.*

I owe enormous debt to my mentors, Peter Godfrey-Smith, Roald Hoffmann, and Philip Kitcher, for their guidance and insight. I have also greatly benefited from the advice, counsel, criticism, and good humor of many colleagues and friends with whom I have discussed the ideas in this book. I thank Jason Alexander, Robert Batterman, Cristina Bicchieri, Nancy Cartwright, John Dupré, Marcus Feldman, Paul Guyer, Paul Griffiths, Tom Griffiths, Gary Hatfield, Elihu Gerson, Murray Goodman, Ronald Giere, Patrick Grim, Volker Grimm, Stephen Hartmann, Rom Harré, Robin Hendry, Paul Humphreys, Ben Kerr, Steven Kimbrough, Rob Kohler, Henrika Kuklick, Richard Levins, Susan Lindee, Elisabeth Lloyd, Tania Lombrozo, Ian Lustick, Uskali Maki, Sergio Martínez, Sandra Mitchell, Mary Morgan, Margaret Morrison, Paul Needham, Joan Roughgarden, Brian Skyrms, Tony Smith, Elliott Sober, Karola Stotz, Paul Teller, Martin Thomson-Jones, Bas van Fraassen, Ward Watt, and Rasmus Winther.

Over the years, I have learned much from audiences at the University of Alabama, Australian National University, University of Bristol, University of Chicago, Cornell University, Duke University, University of Durham, Franklin and Marshall College, Indiana University, the London School of Economics, University of Missouri, University of Pennsylvania, University of Pittsburgh, San Francisco State University, Southern Methodist University, the Sorbonne, St. Louis University, Tilburg University, Universidad Autónoma Metropolitana, University of California at San Diego, Universidad Nacional Autónoma de México, University of Utah, Washington and Lee University, the Philosophy of Science Association, and the International Society for History, Philosophy, and Social Studies of Biology, where I have presented the ideas that form the core of this book.

Thanks are also due to Bill Angeloni and John Kern, who shared their invaluable first-hand knowledge about the Bay model operations, as well as Robert Glass of the National Archives–Pacific Region, for assistance with the Army Corps and John Reber archives.

Much of the first draft of this book was written, and the core ideas worked out, during three stimulating visits to the Australian National University. I thank the University of Pennsylvania, the National Science Foundation (SES-0620887, SES-0957189), and ANU for making these visits possible. Many thanks also to the philosophy of science community at ANU, including Rachael Brown, David Chalmers, Benjamin Fraser, Alan Hájek, Benjamin Jeffares, Aidan Lyon, Kelly Roe, Daniel Stoljar, and especially Brett Calcott, John Matthewson, and Kim Sterelny.

My writing partner Talissa Ford helped me make it through some of the most challenging stretches of writing, and I hope I was able to do the same for her.

This book has been greatly improved by the careful reading and insightful comments of colleagues and students. Many thanks to Elisabeth Camp, Roman Frigg, Peter Godfrey-Smith, Glenn Ierley, Alistair Isaac, Richard Lawrence, Arnon Levy, Jay Odenbaugh, Isabelle Peschard, and Scott Weinstein. And I feel immense gratitude to my students, past and present, Matthew Bateman, Wiebke Deimling, Scott Edgar, Alkistis Elliott-Graves, Paul Franco, Kory Johnson, Daniel Issler, Ryan Muldoon, Emily Parke, Carlos Santana, and Daniel Singer, for their comments on my manuscript and for providing an exemplary philosophical community.

Finally, my deepest thanks are to my wife Deena for her intellectual and editorial good sense, and her immeasurable support and love.

■ Simulation and Similarity

1 Introduction

1.1 ∎ TWO AQUATIC PUZZLES

San Francisco's drinking water travels 167 miles from Yosemite National Park's Hetch Hetchy reservoir to the Pulgas Water Temple in San Mateo. The Temple marks the end of the water's journey with the biblical promise: "I give waters in the wilderness and rivers in the desert, to give drink to my people" (Isaiah 43:20). These words were inscribed to remind the residents of the Bay Area of the fragility of their water supply. This fragility weighed heavily on the minds of many San Franciscans, but on no one more than John Reber, an amateur musical theater producer, whose business cards admonished "without water there is no life."

Reber's concern about the Bay Area's water problems led him to propose an ambitious solution, which would not only supply San Francisco with nearly unlimited drinking water, but also revolutionize the area's transportation, industrial, military, and recreation infrastructure. His plan was to dam up the Bay.

By constructing two massive dams at the entrances to the north and south arms of the Bay, Reber believed that he could capture the 33,000,000 acre-feet of fresh water from the Sacramento and San Joaquin rivers that wash out to sea every year. The tops of his dams would be sealed so that they could hold new highways and rail tracks. And he would backfill 20,000 acres of land, creating new military bases, an airport, and land for industrial and recreational uses (Figure 1.1).

Reber thought that his plan would solve the Bay Area's water and land problems at a single stroke, and he garnered much support from political leaders, retired military brass, and the general public. However, his critics worried that it would destroy commercial fisheries, render the South Bay a brackish cesspool, and create problems for the ports of Oakland, Stockton, and Sacramento (Jackson & Paterson, 1977). Among the opponents of the plan was the Army Corps of Engineers, which was charged with studying the overall impact of the Reber plan and several other plans for building barriers in the Bay. The Corps recognized the benefits that the Reber plan might bring to the area, but it was certain that damming the Bay would have serious, unintended consequences. It recognized that a battle of words would not be helpful in advising regional authorities; it needed hard data. But how could such data be collected without

2 ■ Simulation and Similarity

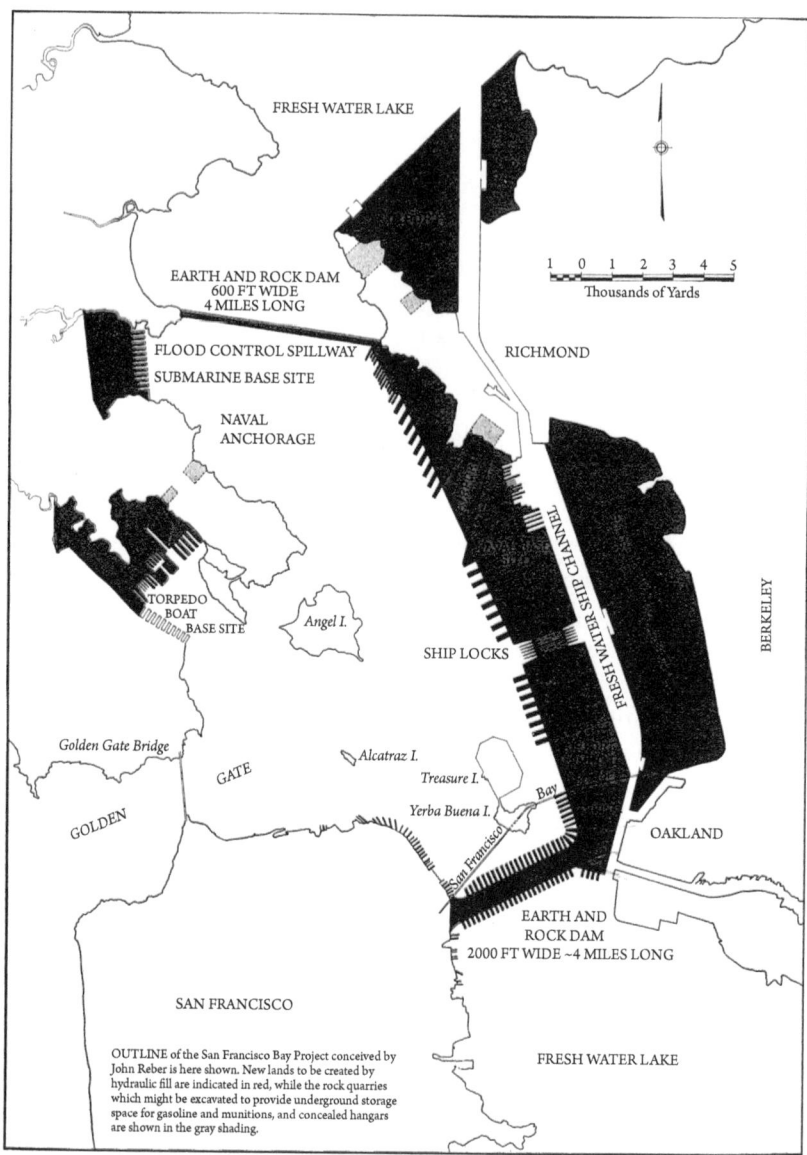

Figure 1.1 A schematic of the Reber plan. No. 77-94-09; "San Francisco Bay Project - The Reber Plan," papers of John Reber relating to the Reber Plan for San Francisco Bay and other subjects, c. 1917–1967; Publications and reports describing and promoting the Reber Plan, National Archives and Records Administration-Pacific Region (San Francisco).

actually building the dams and risking harm to the Bay? Its solution was to build a massive hydraulic scale model of the San Francisco Bay.

This wasn't any ordinary scale model. It was the "San Francisco Bay in a Warehouse" (Huggins & Schultz, 1967), an immense structure, housed in a Sausalito warehouse, that started out at a size of about one acre and has grown to about 1.5 acres today. Hydraulic pumps simulated the action of tidal and river flows in the Bay, modeling tides, currents, and the salinity barrier where fresh and salt water meet (Huggins & Schultz, 1973).

The model was built with Reber's encouragement and political support, and he attended the opening ceremony. But Reber didn't live to see his plan constructed within the model. Several years after his death, the Corps released its final report. It showed that Reber's plan would have disastrous consequences, and could never really generate the freshwater lakes that Reber promised.

A very different sort of model was built to address another aquatic puzzle. After World War I, there was an unusual shortage of aquatic life in the Adriatic Sea off the coast of Italy. This seemed especially strange because fishing had slowed considerably during the war. Most Italians believed that this should have given the natural populations a chance to increase their numbers, but this did not happen. The Italian biologist Umberto D'Ancona was on the case. After carefully analyzing the statistics of fish markets he discovered an interesting fact: The population of sharks, rays, and other predators had increased during the war while the population of squid, several types of cod, and Norwegian lobster had decreased. How could this be? Why did the small amount of fishing associated with the war favor the sharks?

D'Ancona brought this question to his father-in-law, the well-known mathematician and physicist Vito Volterra. Volterra approached the problem not by studying the fishery statistics directly, and not by building a physical model, but by constructing a mathematical model composed of one population of predators and one population of prey (Volterra, 1926a). The result was what we now know as the Lotka–Volterra model of predation, which is described by the following two differential equations:

$$\frac{dV}{dt} = rV - (aV)P \qquad (1.1)$$

$$\frac{dP}{dt} = b(aV)P - mP \qquad (1.2)$$

By analyzing the models described by these equations, Volterra solved the puzzle of the fishery shortages. His model predicted that intense levels of a general biocide, which kills both predators and prey at the same time, would be relatively favorable to the prey, whereas lesser degrees of a biocide would favor the predators. From this he reasoned that heavy fishing, a general biocide, favors the prey and light fishing favors the predator. Because World War I had slowed Adriatic fishing to a trickle, his model suggested that the shark population would be especially prosperous during this time of reduced biocide. This is not something that Volterra or anyone else would have expected *a priori*. However, armed with the dynamics of his mathematical model, Volterra found a solution to this perplexing problem.

These two episodes are paradigm cases of scientists solving problems by *modeling*, the indirect study of real-world systems via the construction and analysis of models. Contemporary literature in philosophy of science has begun to emphasize the practice of modeling, which differs in important respects from other forms of representation and analysis (e.g., Godfrey-Smith, 2006; Wimsatt, 2007; Weisberg, 2007b). This literature has stressed the constructed nature of models (Giere, 1988), their autonomy from theories (Morgan & Morrison, 1999), and the utility of their high degrees of idealization (Levins, 1966; Wimsatt, 1987; Batterman, 2001; Hartmann, 1998; Strevens, 2008; M. Weisberg, 2007a). What this new literature about modeling lacks, however, is a comprehensive account of the models that figure in the practice of modeling. In particular, and most importantly for the current purposes, this literature has not fully explored the role of theorists' intentions in all aspects of modeling, including the individuation of models, the coordination of models to real-world systems, and the evaluation of the goodness of fit between models and the world. The aim of this book is to provide such an account, along with a new analysis of the practice of modeling and the extent to which idealization plays a role in it.

1.2 ■ MODELS OF MODELING

This book is an account of modeling and idealization in modern scientific practice. In addition to mathematical modeling, I will discuss types of modeling that have not traditionally been at the center of the philosophical literature, including concrete modeling, computational modeling, the investigation of specific real-world systems with highly idealized models, and the practice of studying models that may not map on to any particular system in the world.

As many of the examples discussed in this book will show, modeling is not always aimed at purely veridical representation. What is distinctive about the work of the Army Corps, Volterra, and other modelers is that they did not proceed by trying to construct fully accurate representations of their target phenomena. Rather, they worked hard to identify the features of these systems that were most salient to their investigations.

Understanding these types of theoretical practices is the goal of this book. I will begin by distinguishing the three main types of modeling deployed in contemporary science: mathematical modeling, computational modeling, and concrete modeling. This, along with a consideration of what these kinds of models have in common and whether or not there are any other kinds of models, is the subject of Chapter 2.

Chapters 3 and 4 are about the nature of these three types of models. I will argue that models are interpreted structures, and I will discuss what kinds of structures can serve as models and how these structures can differ in their representational capacities. In Chapter 4, I will additionally argue against the claim that all mathematical models should be thought of as fictional scenarios, but I will suggest ways in which the insights of philosophers holding this view can be incorporated into my own account of models.

The next part of the book returns to the theoretical practice of modeling, moving from questions about the nature of models to questions about the practice of modeling. Chapter 5 will focus on the simplest case of modeling, which I call *target-directed modeling*. This is the case where theorists try to explain the properties of a single target using a single model, as in the Bay model case.

The next two chapters consider more complex instances of the practice of modeling. In Chapter 6, I will discuss the practice of idealization, where theorists intentionally distort their models relative to their intended targets. Although the practice of idealization can sometimes seem without constraint, I will develop a framework for thinking about the various goals that idealizations serve and how these goals trade off against one another. Chapter 7 is about modeling without a single, specific target. I will discuss three cases of modeling without a specific target: generalized modeling, hypothetical modeling, and targetless modeling.

Drawing on this expanded account of the practice of modeling, I will next turn to the question of the relationship between models and real-world target systems. As I will show, models are not always veridical; they do not always truthfully describe all aspects of their targets. So what is the relation between models and their targets, especially in cases where the model is highly idealized with respect to the target? I answer this question in Chapter 8. I will begin by discussing some of the influential philosophical accounts of this relation, including isomorphism, homomorphism, and partial isomorphism. I will also consider accounts of similarity developed by psychologists and computer scientists. Drawing on these resources, I will offer my own account of the model–world relation, which I call *weighted feature-matching*.

Throughout the book, little attention will be given to how models are tested with data. However, one of the upshots of Chapters 6 and 7 is that scientists are likely to generate a large number of at least partially incompatible models when dealing with complex phenomena. This requires that they employ *robustness analysis*, a tool that theorists use to determine which parts of their models reliably represent real-world phenomena. Chapter 9 gives my account of robustness analysis and explains how it helps theorists to understand what their models are telling them about the world.

Just as theorists offer incomplete, idealized models of their targets, so must philosophers. Theoretical practice is rich and multilayered, and the world is often uncooperative. Paul Feyerabend's dictum that "anything goes" in science often seems true of theoretical practice. Nevertheless, by developing philosophical accounts of modeling, we can start to get a handle on theoretical practice. But just as in a representation of any other complex phenomenon, philosophical analysis will necessarily be partial and incomplete. Thus the accounts described in this book are themselves models of modeling.

2 Three Kinds of Models

Contemporary scientific practice employs at least three major categories of models: concrete models, mathematical models, and computational models. Roughly speaking, concrete models are physical objects whose physical properties can potentially stand in representational relationships with real-world phenomena. Mathematical models are abstract structures whose properties can potentially stand in relations to mathematical representations of phenomena. Computational models are sets of procedures that can potentially stand in relations to a computational description of the behavior of a system.

Earlier philosophical literature has focused primarily on mathematical models, with occasional mentions of concrete models.[1] However, concrete models still find scientific applications, especially in the applied sciences and engineering. Computational models are rapidly becoming the primary form of scientific model, and they are especially ubiquitous through the biological and social sciences. But mathematical models still dominate the attention of much of theoretical science, as well as the philosophers who study it.

I will begin this chapter by discussing paradigm examples of these three kinds of models. After considering what they do and do not have in common, I will discuss proposals for both expanding and shrinking the list of model types.

2.1 ■ CONCRETE MODEL: THE SAN FRANCISCO BAY–DELTA MODEL

In Chapter 1, I introduced the San Francisco Bay – Delta model which was constructed by the Army Corps of Engineers in 1956 and 1957 to evaluate the feasibility and desirability of the Reber plan. The model is hydraulic; pumps simulate the action of tidal and river flows in the Bay, modeling tides, currents, and the gradient where fresh and salt water meet (Huggins & Schultz, 1973).

Since the San Francisco Bay and Delta system covers a large surface area, but is not particularly deep in most areas, the model was constructed with a horizontal scale of 1:1000 and a vertical scale of 1:100. The model includes

---

1. Notable exceptions include Hesse, 1966, and Sterret, 2005.

the San Francisco Bay itself, San Pablo Bay, Suisun Bay, the confluence of the Sacramento and San Joaquin rivers, and 17 miles of the Pacific Ocean beyond the Golden Gate Bridge. The model is constructed out of precast, lightweight concrete slabs supported at each corner by leveling screws and sealed together with bitumious joint material (Huggins & Schultz, 1967).

The bottom surface of the model was carved according to a 1943 hydrographic survey and additional soundings recorded for the project. It is embedded with thousands of copper strips (currently totaling about 25,000), which simulate the roughness and texture of the sea bed, as well as compensating for the distortions in flow introduced because of the model's scale.

When it is operating, the model is filled with salt water and a pumping system is used to simulate the tidal cycle. Additional freshwater pipes around the perimeter of the model simulate river and stream flows into the Bay. The salt water is prepared to match the average salinity of the ocean water which enters the San Francisco Bay through the Golden Gate.

In addition to spatial scaling, the mode is also scaled in a number of other respects. Slope and velocity are represented as 10:1, discharge rates 1:1,000,000, and volume 1:100,000,000. The Bay tidal cycle of 24 hours and 50 minutes takes place in 14.9 minutes (1:100 scale) (Army Corps of Engineers, 1981).

Before the model could be used to evaluate the Reber plan, it had to be calibrated to the hydrodynamics of the San Francisco Bay. Given the complexity of the model, this was no small task, and was made even more complicated because there is no single, obvious target for the model to meet. Was it intended to represent the Bay during an average dry season? A dry season of a dry year? A wet season of a wet year? Of course, a perfect model of the Bay, the ocean, the weather, and everything else might be able to adapt to represent all of these conditions, but the scope of this model was narrower. It needed to be calibrated to study the Reber plan's impact on the Bay. In the end, two particular tidal periods were chosen as representative targets.

The primary calibration of the model relied on tidal data collected from 24 tide gauges located throughout the Bay that had direct, scale counterparts in the model. These gauges provided information about the tidal elevation throughout the ebbs and floods of the tide. Additional data were required in order to calibrate the model so it could accurately represent local water velocity, direction, salinity, and sediment levels. The following is a description of how reference data were collected:

> At the time of the autumnal equinox, on September 21, 1956, 15 small craft left the Engineer's Base Yard at Sausalito. Eleven of them took up predetermined fixed positions (control stations) in the bay, to record conditions during the 48-hour period of mean tide, throughout the 550 square miles of bay. ... Eighty

men working in shifts obtained hourly water samples throughout the vertical from surface to bottom, for subsequent determination of salinity fluctuations and suspended sediment concentrations. Current velocities and directions throughout the depth were measured at half-hour intervals ... Since tidal elevations and amplitudes and current velocities vary from day to day, it was essential to make simultaneous measurements for at least one tidal cycle, and therefore, the automatic tide recorders were in operation at the time the other measurements were made. (Huggins & Schultz, 1967, 11)

It would take another 14 months before the Bay model could be fully calibrated to the data obtained on the 21st and 22nd of September. Calibration involved the adjustment of many parameters. The primary inflow and outflow could be calibrated by adjusting two water pumps, which simulated the tide coming in under the Golden Gate Bridge. But there are considerable differences in the height of high tide at different parts of the Bay. This was accounted for both by ensuring that the depth of the model was contoured to proper scale and by changing the simulated bottom roughness by adjusting the copper strips. Freshwater inflows, which simulated streams, were also adjusted to capture salinity variation throughout the model. The model's calibration was ultimately checked by comparing it to further cross-section measurements.

Armed with this high-fidelity model of the San Francisco Bay, the Army Corps of Engineers was now in a position to evaluate the Reber plan. Reber's proposal called for a 600 foot wide, 4 mile long earth and rock dam that stretched from San Quentin to Richmond, and a second barrier 2000 feet wide and 4 miles long, just south of the Bay Bridge, connecting San Francisco to Oakland. The Army Corps studied this proposal by building scaled versions of Reber's barriers, adding these to the Bay model, and measuring the changes in current, salinity, and tidal cycles.

As the Corps predicted, the model showed that Reber's plan would have disastrous consequences for the Bay and its ecosystem. Far from creating the freshwater lakes envisioned by Reber, the barriers would actually create stagnant pools, without the circulation required to maintain healthy aquatic ecosystems. The possibility of introducing openings in the barriers to allow for water flow was also explored. This would not, of course, generate freshwater lakes, but it was thought that this might allow the dams to be built without creating stagnant water behind them. The model showed that building such porous barriers was also a very bad idea because doing so introduced extremely high-velocity currents in the Bay. These would disrupt the normal ecosystem and make navigation in the central part of the Bay hazardous. These Corps was thus in a strong position to denounce Reber's plan on the basis of model-derived data (Army Corps of Engineers, 1963a).

2.2 ■ MATHEMATICAL MODEL: LOTKA–VOLTERRA MODEL

Predation is of great interest to ecologists because it often represents a force that keeps populations below their environment's carrying capacities.[2] It is also a factor which can account for oscillation and other periodic dynamics of populations in which there is no external stimulation, such as in unchanging environments (Ricklefs & Miller, 2000). Theoretical ecologists are interested in studying how predation leads to these phenomena. They construct models to study the factors that control the maximum population size as well as the phase, amplitude, and frequency of oscillations in populations.

Vito Volterra (1926a, 1926b) and Alfred Lotka (1956) independently proposed what we now call the Lotka–Volterra model, a very simple model of predation. Volterra was explicit about his grounds for constructing such a simple model that necessarily excludes much biological complexity. He explained, "Although, at least in the beginning, the representation is very rough ... [I will] schematize the phenomenon by isolating those actions that we intend to examine, supposing that they take place alone, and by neglecting other actions"(Volterra, 1926b, translation G. Sillari).

Lotka and Volterra modeled predation as a population-level phenomenon, so the primary quantity that their model tracked was population density or species abundance. Because these populations interact, their population dynamics are coupled together in the following way: The predators decrease the population of prey by eating them, while the prey increase the population of predators by providing food. Abstractly, the relationship is one of negative feedback. Predators are negatively coupled to the prey, but prey is positively coupled to the predators (Maynard Smith, 1974).

In order to construct simple, population-level models of predation, there are six basic factors to work with: the predator growth and death rates, the prey growth and death rates, the effect of predation on the population of prey, and the effect of prey capture on the population of predators. If we set up the model in terms of rates of increase and decrease, we can collapse intrinsic growth and death rates of the populations into a single growth rate for the prey and, a bit less realistically, a single death rate for the predators. This will give us four factors to track.

Let V stand for the size of the prey population, P for the size of the predator population, and t for time. If we express the basic relationships outlined above with coupled differential equations then we get the following basic

---

2. For a comprehensive review of the classic literature, see Royama, 1971. For more contemporary discussions including the history of predator–prey modeling, see Berryman, 1992; Hanski, Henttonen, Korpimaki, Oksanen, & Turchin, 2001; Briggs & Hoopes, 2004; Jurrell, 2005.

equations:

$$\frac{dV}{dt} = [\text{prey birth rate}] - [\text{prey capture rate per predator}] \quad (2.1)$$

$$\frac{dP}{dt} = [\text{predator births per capture}] - [\text{predator death rate}] \quad (2.2)$$

(after Roughgarden, 1979)

These equations provide a template for a large but tightly linked family of models. Starting from the simple possibilities, the prey growth rate could be linear, exponential, geometric, or logistic. The most typical death rate of the predators in predation models is constant, corresponding to an exponential decay in the absence of prey. More complicated rate expressions are also possible, including functional dependence on environmental parameters and logistic decay when multiple sources of food are present.

Of greater biological interest, at least when considering predator–prey interactions, are the second term in equation (2.1) and the first term in equation (2.2), called the *functional response* and *numerical response* respectively (Holling, 1959). As we can see from the equation template, the functional response is a rate, specifically, the rate of prey capture per predator. The simplest possible assumption is that the functional response is linear, or that the number of prey capture increases linearly with increasing numbers of prey.

Finally, the numerical response term correlates predator births to the number of prey captured. Because of this, the numerical response is itself a function of the functional response. Specifically, the numerical response depends on how many prey are in the population, how good the predators are at capturing them, and how much energy from prey captures can be used to generate new offspring. Naturally, these are complex interactions and will further depend on other environmental variables, such as other stresses on the predator population and energetic cost of offspring. Ecologists almost always collapse most of this complexity into a single parameter and represent the numerical response as a constant multiplied by the functional response.

Now that we have considered how the basic template could in principle be filled in, let's return to the Lotka–Volterra model itself and to Volterra's justification of it. The Lotka–Volterra model is the simplest way to fill in (2.1) and (2.2) because it uses the simplest functions.

Let r stand for the growth rate of the prey population and m stand for the death rate of the predators. The functional response is linear, expressed as a constant a multiplied by V. Similarly, the numerical response is a linear function of the functional response so the whole numerical response expression can be written as a parameter b multiplied by the functional response, or $b(aV)$.

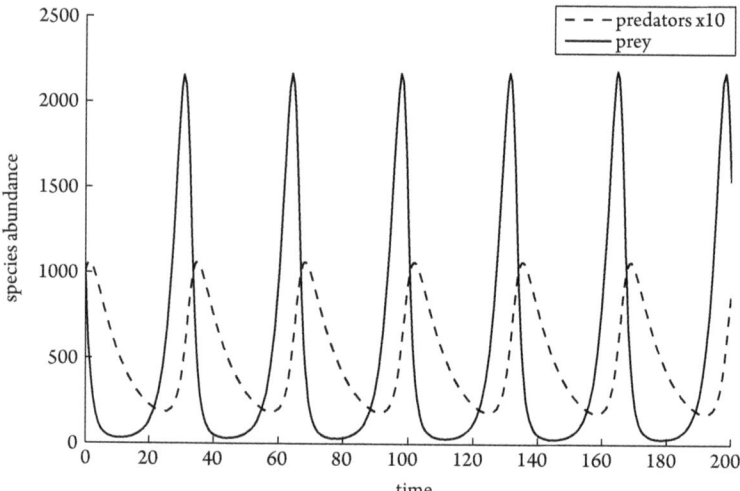

Figure 2.1 Lotka–Volterra model's oscillation

The Lotka–Volterra model is thus described with the following differential equations:

$$\frac{dV}{dt} = rV - (aV)P \tag{2.3}$$

$$\frac{dP}{dt} = b(aV)P - mP \tag{2.4}$$

These equations describe a model with a dramatic result: the predator and prey populations will oscillate indefinitely, out of phase with one another (see Figure 2.1). Although for every set of parameter values with species coexistence there exists one equilibrium where the populations do not oscillate, this equilibrium is unstable and hence the model populations will continue to oscillate if they are perturbed even slightly off these equilibrium values.

The most important property of the Lotka–Volterra model is the *Volterra property*, the key component of the *Volterra principle* (Weisberg & Reisman, 2008). The Volterra property states that a general biocide, any substance which has a harmful effect on both predators and prey, will *increase* the relative abundance of the prey population. To see this, we first need to solve for the equilibrium abundances of the species by setting each differential equation to zero. After some algebra, we find that the equilibrium values are:

$$\hat{V} = \frac{m}{ab} \tag{2.5}$$

$$\hat{P} = \frac{r}{a} \tag{2.6}$$

These are unstable equilibria, but they correspond to the average abundance of the predator and prey species over indefinitely long time periods.

We can derive the Volterra property from the Lotka–Volterra model by first expressing the ratio of the average size of the predator population to the average size of the prey population ($\frac{\bar{P}}{\bar{V}}$) as ρ. Decreases in ρ will correspond to increases in the relative size of the prey population.

From equations (2.6) and (2.5) we can see that

$$\rho = \frac{rb}{m} \qquad (2.7)$$

The next step is to consider how a general biocide affects the model populations. We can represent the introduction of a biocide as corresponding to changes in r and m. Specifically, biocides decrease the prey growth rate (r) and increase the predator death rate (m). Inspecting ρ, the expression for the ratio of average densities, we can see that ρ(biocide) < ρ(normal) (May, 2001; Roughgarden, 1979, 439). Since smaller values for ρ mean a larger relative size of the prey population, the population of prey will increase relative to the number of predators when a biocide is applied. This demonstrates the Volterra Property: a general biocide will increase the relative size of the prey population. Since heavy fishing is a general biocide, this predicts that periods of heavy fishing will favor the prey species. Light fishing, such as during World War I, will favor the predator species.

2.3 ■ COMPUTATIONAL MODEL: SCHELLING'S SEGREGATION MODEL

Thomas Schelling's *Micromotives and Macrobehavior* (1978) contains a very simple model that shows how racial segregation can occur even when no racism is present. This model is agent-based, meaning that each individual is explicitly represented. In his particular implementation of the model, dimes and nickels represented two types of individuals, A and B, and the squares on a chessboard represented spatial location. In this model, each individual prefers that at least 30% of its neighbors be of the same type. So the As want at least 30% of their neighbors to be As and likewise for the Bs. Schelling's neighborhoods were defined as standard Moore neighborhoods, a set of nine adjacent grid elements. An agent standing on some grid element e can have anywhere from zero to eight neighbors in the adjoining elements.

Although Schelling didn't explicitly provide a utility function, the preference described above is usually interpreted to mean that each agent is indifferent between having 30–100% of its neighbors be alike, but will be dissatisfied with fewer than 30% alike. Because of the constraints of the grid's geometry, in the

| initial distribution | $t = 1$ | $t = 2$ | $t = 3$ | $t = 14$ (equilibrium) |

Figure 2.2 An example of Schelling's model of segregation on a 51×51 grid with 2000 agents. Each agent prefers 30% of its Moore neighbors to be the same shape and color. The initial distribution of agents was random, and the model equilibrated after fourteen time steps.

case of a full neighborhood, the preference boils down to wanting to have at least three of one's eight neighbors be alike and to equally prefer 3–8.

The dynamics of Schelling's model involve agents sequentially choosing to remain in place or move to a new location. When it is an agent's turn to make a decision, it determines whether there is a sufficient ratio of alike agents among its neighbors. If this condition is met, the agent is satisfied and remains where it is. If it is not, the agent then moves to the nearest empty location. This sequence of decisions continues until all of the agents are happy where they are and do not try to move.

The dynamics of the model unfold as a cascade and are depicted in Figure 2.2. Agents that were originally satisfied can become dissatisfied as soon as a neighbor leaves or a new one moves in, and this leads to the new agents becoming dissatisfied. A small patch of dissatisfaction will result in widespread movement, and, ultimately, segregation. Though there are a few grid configurations that are integrated where every agent is happy, these are rare, and nearly impossible to result from agent movement. The equilibrium state of the model is segregation. Thus, Schelling's major result is that small preferences for similarity can lead to massive segregation. This result is quite robust across many changes to the model, including different utility functions, different rules for updating, different neighborhood sizes, and different spatial configurations. In fact, it is extremely hard to avoid segregation when agents have some preference for like neighbors.

2.4 ■ COMMON FEATURES OF THESE MODELS

There are considerable differences between the San Francisco Bay model, the Lotka–Volterra model, and the Schelling model. The Bay model is a major feat of engineering, confined to a specific location in time and space, and requiring careful scientific measurements to learn anything about the target system. On the other hand, the Lotka–Volterra model requires nothing more than paper and a pencil and some knowledge of calculus to learn about its major results. Since it is a mathematical object, it can't be assessed or measured directly, so what we learn about it comes from manipulating the equations that describe it. Schelling's segregation model is similar to the Lotka–Volterra model in that it can be instantiated anywhere, but its overall behavior cannot be analyzed easily. In order to understand its properties, the procedure must be followed—either in physical space the way Schelling did it, or on a computer—in order to learn about its properties of interest. So what do these models have in common such that we can develop a unified account of them?

At their core, all three models consist of an *interpreted structure* that can be used to represent a real or imagined phenomenon. They all have a structure: The Bay model is a concrete tub with a set of pumps and appurtenances, the Lotka–Volterra model is a set of points in a mathematically constructed space, and the Schelling model is a set of states and transitions. What makes these models of the San Francisco Bay, of populations of organisms, and of people in cities lies in the second part of the definition: interpretation. Theorists have to intend a concrete tub to represent the SF Bay, or the state transitions in a computational space to represent movements in the city. These interpretations tell us what the model is about and set up the relations of denotation between models and their intended targets.[3]

2.5 ONLY THREE TYPES OF MODELS?

A number of philosophers (e.g., Downes, 1992; Winther, 2006) have criticized the classical literature on models and modeling because it has been too narrowly focused on a single type of model: dynamical mathematical models represented by differential equations. They argue that not every model is mathematical and that even mathematical models have more diversity than classical accounts of modeling allow for. I am very sympathetic to this criticism and think that philosophers of science need to make room for computational and concrete models in their accounts of modeling. However, my three-way categorization is stricter than some of the other ones on offer. Is this categorization scheme to narrow?

While I am not certain that I have given a closed, exhaustive list, I believe that my three types of models can accommodate the exemplars commonly discussed in the philosophical literature. Beyond the kinds of cases I have discussed, there

---

3. My view of models as structure plus interpretation has many affinities to the work of Giere (1988), Cartwright (1983), and Teller (2001). My view departs in significant ways from the new structualists such as Ladyman (1998), French (2006), and da Costa & French (2003), but we are still fellow travelers in emphasizing that comparing structure to structure is at the core of modeling. For further discussion and similar views see also Godfrey-Smith (2006), Lloyd (1994), van Fraassen (1980), Humphreys (2007), Hughes (1997), and Pincock (2011).

are three categories that are commonly mentioned by philosophers of science: model organisms, verbal models, and idealized exemplars. Let's consider each of these cases in turn.

Model organisms are specific species that are used by biologists to study biological phenomena that are not necessarily confined to those particular species. In genetics, *Drosophila melanogaster* and *Escherichia coli* are extremely common model organisms. Medical researchers often use dogs to serve as models of human beings. And psychologists have long used rodents as models of learning across species. Usually the particular organisms used in laboratory experiments are highly inbred, pure strains not found in their current form in the wild.

Model organisms are deployed for several purposes. Sometimes, scientists use them when studying a phenomenon requires investigating some organism or other, but not any particular organism. For example, biologists interested in gene regulation often study yeast. Brewing and baking aside, yeast gene regulation isn't intrinsically interesting. Rather, these organisms are studied because they can be easily handled in the laboratory and the biochemistry of gene regulation is thought to be similar across all of life.

Another use of model organisms is in cases where a particular class of organisms is being studied, and biologists need an exemplar of this class. For example, ecologists studying invasive species have often examined the Australian rabbit invasion (Scanlan, Berman, & Grant, 2006) or the kudzu plant (Forseth & Innis, 2004). Still another use of model organisms comes when scientists want to know about a particular organism, say humans, but ethical or monetary considerations prevent the experiments being carried out on the intended target.

One of the foci of the literature about model organisms concerns the target of such models. If we consider the kudzu plant as a model organism, then its target is a very wide range of actual and possible invasive plant species. We are supposed to be able to generalize about what we learn from kudzu in order to understand and prevent future invasions. In addition, model organisms seem distinct from other paradigm cases in that we learn about them not by doing mathematical analysis, but by performing empirical experiments (Ankeny, 2001; Griesemer & Wade, 1988; Griesemer, 2003).

Model organisms are obviously an important class of models, but I think that they can be accommodated in the framework I have presented thus far because model organisms are concrete models. Although they are not constructed, like the San Francisco Bay model, they are concrete systems that resemble concrete targets. The special properties of model organisms discussed above are not unique to model organisms, but are ubiquitous features of modeling practice. For example, as I will discuss in Chapter 7, many models have generalized target systems. Even the Bay model, which was calibrated to the tides of two specific days, had to stand in as a general model of the Bay – Delta system. In this way, it is like the case of a single species being used to study a larger group of species.

In addition, other kinds of concrete models besides model organisms are studied by doing experiments. As I discussed in Section 2.1, the San Francisco Bay model was studied by manipulation; model dams from the Reber plan were inserted to see the effect of blocking the ocean from entering most of the Bay. So in many ways, the Bay model is analogous to model organisms. Until synthetic biology matures, the main difference between the Bay model and model organisms is that one is constructed and the other has its origin in the wild. Nevertheless, both are concrete models.

Another candidate class of models is what some philosophers have called *verbal models* or *narrative models* (Winther, 2006). Such models are presented in narrative form, but usually involve verbally sketching some possible mechanism and explaining how the behavior of this mechanism could account for some real-world phenomenon. For example, Shepard and Metzler (1971) gave subjects pairs of pictures of geometric shapes in different orientations. These shapes were either identical or nonsuperimposable mirror images of one another. Subjects needed to determine whether they were identical or not and press a button indicating their choice. They found that the amount of time it takes subjects to determine whether or not the shapes are identical varies as a linear function of the angle of rotation of the objects.

In order to explain this finding, Shepard and Metzler suggested that subjects might be mentally representing the pictures as three dimensional shapes and actually rotating them mentally. We can think of this explanation as involving the introduction of a model in that Shepard and Metzler propose a mental mechanism, think through its consequences, and then hypothesize that it would explain the results that they obtained. However, no mathematics is used to formulate this model and nothing concrete was constructed. Hence some have called models like this "purely verbal."

I am happy to grant that Shepard and Metzler formulated a model to explain their experimental result, but I don't think this represents a distinct category of modeling. First, how a model is described is distinct from what the model is. As I will discuss in Chapter 3, descriptions of all three types of models can be verbal, mathematical, diagrammatic, and so forth. The fact that Shepard and Metzler use words, not equations, to describe their model doesn't give it a special status.

In addition, I believe that the reason verbal models seem to be a special category is because they are often underdescribed. In their original paper, Shepard and Metzler don't say much more about the mental rotation mechanism then what I have described above. But that falls short of fully describing a model. If we consider a further elaboration of the model, we will get more insight into the model's character.

I suspect that an elaboration of Shepard and Metzler's model would contain a lot of mechanistic detail of the sort that visual input V triggers mental mechanism M, which is processed by P, and gives output O. When spelled out in detail, the verbal model becomes a verbal description of a computational model, a procedure that explains the phenomenon. Thus, verbal models, at least the sort one commonly encounters in psychology, are often verbally described computational models.

The final type of case that is neither concrete nor obviously mathematical is an *idealized exemplar*. One such case is discussed by Steve Downes in a landmark paper about the variety of types of models: the standard model of a eukaryotic cell.

> ... consider a typical biology textbook drawing of a cell. In most texts a schematized cell is presented that contains a nucleus, a cell membrane, mitochondria, a Golgi body, endoplasmic reticulum and so on. In a botany text the schematized cell will contain chloroplasts and an outer cell wall, whilst in a zoology text it will not include these items. The cell is a model in a large group of inter-related models that enable us to understand the operations of all cells. The model is not a nerve cell, nor is it a muscle cell, nor a pancreatic cell, it stands for all of these. (Downes, 1992, 145)

In order to assess whether idealized exemplars are a separate category of models, we should separate some of the bases on which such a claim could be made. According to Downes, the standard model of the cell is: (1) presented diagrammatically, (2) further developed verbally, (3) abstract relative to any real cell, and (4) idealized relative to any real cell. Let's consider these claims in turn.

The fact that the standard model of the cell is presented diagrammatically does not make it a special category any more than if it were presented verbally. How a model is presented involves the properties of the model's description, not the model itself. Concrete models can be presented mathematically, and mathematical models can be presented concretely. For similar reasons, I don't see why the fact that this model is developed verbally makes it different from any other kind of model.

I suspect that when philosophers emphasize the verbal or narrative form that standard models of the cell take, they are really pointing to something more like the richness of these descriptions. The model presented in textbooks isn't sparse like the Lotka–Volterra model. Rather, it is more like a concrete model in the amount of detail that it conveys.

Downes points out that the textbook model of the cell is both abstract and idealized relative to any real cell. It is abstract because it isn't a model of any particular kind of cell; it is a model of properties shared by all eukaryotic cells. Relatedly, it is idealized because its generality forces some parts of the model to be distorted relative to any real cell. I think these are both interesting properties, and are especially vivid in the cell case, but they are really quite ubiquitous across all sorts of models. Certainly all three of my main examples are both abstract and idealized relative to their targets.

If there is nothing about the class of idealized exemplars that makes them a unique case worthy of separate treatment, what are they? I think that the standard model of the eukaryotic cell is actually a concrete model, albeit one that probably has never been built. Descriptions and diagrams in books are both aids for the hypothetical construction of this model and guides for thinking about it in the absence of its construction.

I have argued that model organisms and idealized exemplars are actually types of concrete models, and that verbal models are actually model descriptions specifying any of the three types of models. I cannot claim to have exhausted all possible types of models and I think it would be foolish to claim that it is impossible that there is another class of models that doesn't fall into my three-way distinction. However, all of the examples of models and modeling commonly discussed in the philosophical literature can be thought of as concrete, mathematical, or computational, or possibly as some hybrid of the three.

2.6 ■ FEWER THAN THREE TYPES OF MODEL?

In the last section, I argued that there aren't good reasons to proliferate the list of model types beyond three. Now I will consider the converse question: are there actually fewer than three types of models?

The most obvious way to reduce the number of model types is to argue that computational models are just a special case of mathematical models. After all, computational operations are at base mathematical operations. They are formally described in terms of states and transitions, and discrete mathematics gives a theory of these transitions. Two other possibilities are that we construe concrete models as somehow being mathematical, or mathematical models as being concrete, perhaps in virtue of their really being fictional scenarios.

First, when we ask "ultimately, how many types of models are there?" we can be asking at least three different questions. First, we might be asking about the face-value practice of science. This would require us to ask, "How many types of models do scientists talk about?" Second, we could be asking about ontology. We could ask, "Fundamentally, what are models, and how many types of things are they?" Finally, we could be asking a question like, "In order to build an account of the practice of model-based theorizing, how many categories of model will we need?" This last category of question is what Stacy Friend has called the *epistemic* level of philosophical theorizing (Friend, 2009; see also French, 2010).

Throughout the book, I will comment on all three types of questions, but my primary concern in this chapter and throughout the book is with the epistemic level. When I say that there are three types of models, I am not making a purely descriptive claim about how many categories of models are recognized by scientists, nor am I making a point about fundamental ontology. Rather, I am arguing that a philosophical account of models and modeling needs these three categories to account for modeling as it is practiced in contemporary science.

Of course, appeals to the needs of a practice-based account cannot completely settle this issue because much depends on what one hopes to do with such an account. Some philosophers of science are interested in a very fine-grained characterization of the kinds of tasks that can be accomplished with different models. For them, it would be very important to show how two models I would consider of the same type—model organisms and the Bay model—differ from one another. My goals for this book are different. I want to understand the representational capacities of broad classes of models, the role of theorists' intentions in the individuation and interpretation of models, the coordination of models to real-world systems, and the evaluation of the goodness of fit between models and the world. In light of these goals, let's consider some of the possibilities for collapsing the categories that I raised above.

First, why can't we simply describe computational models as mathematical models? The answer is simple: We can. However, I think it is important to distinguish between these mathematical models and computational models because of how they are used in giving scientific explanations. When one invokes a computational model to explain some phenomenon, one is typically using transition rules or algorithm as the *explanans*. Schelling explained segregation by pointing out that small decisions reflecting small amounts of bias will aggregate to massively segregated demographics. Neither the time sequence of the model's states nor the final, equilibrium state of the model carries the explanatory force; the algorithm itself is needed. Conversely, in the types of models I am calling mathematical models, the mathematical structure, or relationship among variables, carries the explanatory power. So while ontologically they are not distinct types of models, they function differently in practice and have different representational capacities, so they should be given their own categories.

Another possibility is that concrete models are really mathematical or computational models. Although I don't know of a philosopher who defends this view explicitly, some defenders of structural realism have contended that all physical objects are ultimately mathematical structures (French & Ladyman, 2003). A related, but less extreme view is that, while concrete models are not mathematical, their scientific value as potential representational devices often derives from their mathematical properties. Pincock (2011, §5.3) has recently defended this view.

Some evidence in favor of these structuralist, "maths as central" views can be found by considering the indispensability of mathematics to all types of modeling. Pincock points out that even when concrete structures are deployed in modeling, comparison of these models to target systems very often requires mathematics. When developing the San Francisco Bay model, engineers calculated certain dimensionless quantities for both the target and the model, and then compared these quantities in order to ascertain the similarity relationship between the model and the target (Army Corps of Engineers, 1963a, 1963b; see also Kline, 1986; Batterman, 2001, for more general discussions of the use of dimensionless quantities).

It is unclear how general Pincock's observation really is, though, as there are parts of biology and chemistry that seem to compare models to targets in structural, behavioral, and other nonobviously mathematical ways. But even if we grant this point in favor of the family of structuralist views, I come back to the point I made above about computational models. Independent of the ultimate ontology of concrete models, we can see that they function differently in scientific practice from mathematical and computational models. For one thing, they can be studied directly, and do not require the mediation of equations.

Further, concrete models are much richer in properties than just the mathematical ones. Even if a Pincock-like point can be made that all or most of the important uses of the model have to proceed by using the mathematical properties of the model, it is not clear that these can be decoupled from the rest of the model. For example, while the Bay model engineers relied on dimensionless quantities to verify the model's similarity to its target, the similarity was assessed between all of the model's hydraulic properties and the target, not just those mathematical properties. So I think we need to keep concrete models as a distinct category.

A final proposal for reducing the number of types of models was developed by Martin Thomson-Jones. He argues for a unification of all types of scientific models by claiming that such models are best thought of as propositions. A propositional model "is a set of propositions, the members of which together represent some system from the relevant domain of inquiry as having certain features, behaving in certain ways, and so on" (Thomson-Jones, 1997, 11). Thomson-Jones argues that this notion of a model can make sense of mathematical models and also of several other hard cases including the aforementioned model of the eukaryotic cell and classes of models in physics such as the Bohr model of the hydrogen atom.

Although I am interested in Thomson-Jones' account because of his claims about unification, his own motivation for proposing that models are sets of propositions was to deal with a certain class of models. These models are extremely general descriptions of actual and potential systems and seem to differ from mathematical models. Discussing the Bohr model, Thomson-Jones says that

> ...the Bohr model of the hydrogen atom is not a representation of any *particular* evolution a hydrogen atom might undergo. Instead, it contains a significant amount of modal information about hydrogen atoms: the electron can move around the proton in any circular orbit which satisfies the constraints laid out in the course of presenting the model [...]. Nothing else is permitted. (33)

Thomson-Jones argues that since mathematical models are supposed to be single trajectories through state space corresponding to the historical evolution of a single system, the Bohr model cannot be a mathematical model.[4] He similarly argues that cell models contain information about space and cellular processes, none of which can be understood as corresponding to a trajectory through state space.

Since Thomson-Jones thinks that it is awkward to treat these cases as examples of mathematical models, he proposes that these kinds of cases, along with mathematical models, are best thought of as sets of propositions. For example, the Bohr model would be thought of propositions such as:

1. Hydrogen atoms are composed of one positively charged proton and one negatively charged electron.
2. The electron orbits the proton in a circular orbit and has angular momentum $n\hbar$, where n is a positive integer.

The relevant propositions would thus involve mentioning physical objects and properties like hydrogen, electrons, and negative charge, and this would explain how the model can have representational content. A very similar analysis could be extended both to fully nonmathematical cases like models of the cell and to more heavily mathematical cases. Thus, Thomson-Jones argues, we have a unified account about the nature of models: they are sets of propositions.

My response to Thomson-Jones's alleged unification has two parts. First, I think it is fairly obvious that concrete models like the San Francisco Bay model are not sets of propositions. Sets of propositions can describe such models, but the models themselves are concrete, physical objects. In fairness to Thomson-Jones, he didn't talk about concrete models in his article, but if we take them seriously, then I don't think we can achieve full unification of the three types of models using propositions.

The second part of my response is more sympathetic. Earlier in this chapter, I argued that what the three types of models had in common was structure and

---

4. I think this move is too quick, because many proponents of state-space approaches to mathematical models, such as van Fraassen and Lloyd, would be happy to say that *sets* of trajectories through the state space constitute the model.

interpretation. Considered at the right level of abstraction, interpreted structures can be thought of as propositions. They are semantic contents which potentially represent their targets. However, this is a relatively weak claim. All it says is that a model has the same kind of representational content as other bearers of content such as sentences and pictures. This does not commit us to an ontology of propositions.

In addition, I do not think that it is sufficient to say that all models are unified because they can be thought of as semantic content that potentially represents real-world phenomena; this is just the same as saying that different types of models are all models. In order to understand the practice of modeling, including the nature of models and model–world relations, we should be thinking about models much closer to the level of abstraction at which they function. If, for example, theorists learn about segregation in real cities by asserting similarity relations between a procedure instantiated by model residents and the thought processes of actual residents, then we need to individuate models at this level of abstraction.

Thus, we are left with the view that there are three types of models, and each of these is composed of a structure along with an interpretation of that structure.

3 The Anatomy of Models

In the last chapter, I argued that scientific models are composed of a structure and a scientist's interpretation of that structure. The purpose of this chapter is to explore this dual-aspect account in detail.

3.1 ■ STRUCTURE

Concrete, mathematical, and computational models differ with respect to their structures. In this section, I will discuss the structures at the heart of each type of model.

3.1.1 Concrete Structures

Concrete models are real, physical objects that are intended to stand in representational relationship to some real or imagined system, set of systems, or generalized phenomenon. I will call the systems to which a theorist intends her model to apply the *intended targets* of the model.

The Bay model is a paradigm case of a constructed, concrete model. Its target is, of course, the San Francisco Bay. Other historically important examples include ancient Greek models of the planets, Maxwell's mechanical model of the ether, Watson's and Crick's model of the structure of DNA, and scale models of airplane wings and engines. Some of these models, such as the Bay and DNA models, stood in successful representational relations. Others, such as the Greek models of the planets and Maxwell's ether models, failed to resemble their intended targets because those target systems do not exist.

In addition to models which are literally constructed, scientists can also work with naturally occurring concrete models, which are structures and phenomena that already exist in nature and resemble other phenomena of interest. Perhaps the most widely used natural models are model organisms (Griesemer & Wade, 1988; Griesemer, 2003; Weber, 2005; Winther, 2006). Fruit flies, for example, are often called the "test tubes of molecular biology" because of their ubiquity and utility in genetics. For mammalian studies, especially those involving medical research, mice, rats, dogs, and nonhuman primates can all be studied in place of studying humans. Of course, fruit-fly molecular biology is not the same as the biology of all other animals, and mice, while similar to human beings in some ways, are obviously different in many others. But for particular research purposes, these natural models share enough properties with their intended targets to make their study scientifically fruitful.

A more complex, but substantially similar, case involves the use of natural experiments in population dynamics, geology, and climatology (Richardson, 2006). In this case, there isn't a particular object or organism that stands in for another organism or class of organisms. Rather, a dynamic phenomenon taking place in time becomes a model for targets that are inaccessible temporally or spatially. High-pressure water quickly diffusing through rocks in one place might serve as a useful model for low-pressure, long-diffusing water in another. While the relationship between these models and theorists' targets are very similar to natural models, far more emphasis is put on behavioral similarity than on structural similarity in these cases.

Finally, a variant of the natural experiments studied by geologists is natural population experiments studied by anthropologists. For example, Jared Diamond (1999) argues that geographical factors and the availability of crops and animals for domestication are two of the major factors which determine how effectively one population can dominate another when they come in contact. Diamond's goal in developing this theory is to explain the success that Europeans had in conquering the Americas and Africa. These interactions were so complex and took place on such a large scale that the most effective strategy for determining the operating mechanisms was to study small-scale, regional conflicts. The Maori people from northern New Zealand, for example, nearly exterminated the Moriori from the Chatham Islands when the former invaded. The Moriori had reverted to a hunter-gather lifestyle because Polynesian tropical crops could not grow in the Chathams' cold climate, while the Maori had a relatively advanced agricultural society, which allowed for military specialization. The Maori and Moriori case illustrates some of the causal factors at play in the larger-scale questions discussed in Diamond's book. Like the Army Corps studying the Reber plan's consequences with the Bay model, this natural experiment allowed more direct comprehension of the operant causes without trying to extract them directly from the intended target. In other words, the natural experiment itself is a structure that can serve as a model.

What we have learned from these examples, and those discussed in Chapter 2, is that concrete models can take three forms: They can be literally constructed structures, they can be pre-existing structures, or they can be structures that are described but never built. In each case, a concrete structure becomes a model when it is intended to be used as a model.

3.1.2 Mathematical Structures

The second type of scientific model is a mathematical model, which has been the focus of most philosophical literature about models and modeling. In fact, this literature has focused on a very small subset of these models: dynamical models of the sort described by differential equations.

The traditional philosophical account of mathematical models says that such models are either set-theoretic predicates or sets of trajectories in a state space. For example, in Suppes' original development of the model-based *semantic view of theories*, he argued that mathematical models are very similar to the notion of a model that comes from mathematical logic. On this view, models are "a certain kind of ordered tuple consisting of a set of objects and relations and operations on these objects" (Suppes, 1960a, 290). He gives a specific example of a model in classical mechanics.

> We may axiomatize classical particle mechanics in terms of the five primitive notions of a set P of particles, an interval T of real numbers corresponding to elapsed times, a position function s defined on the Cartesian product of the set of particles and the time interval, a mass function m defined on the set of particles, and a force function f defined on the Cartesian product of the set of particles, the time interval and the set of positive integers (the set of positive integers enters into the definition of the force function simply in order to provide a method of naming the forces). A possible realization of the axioms of classical particle mechanics, that is, of the theory of classical particle mechanics, is then an ordered quintuple $B = \langle P, T, s, m, f \rangle$. A model of classical particle mechanics is such an ordered quintuple. (291)

Suppes believed that showing the near equivalence of the two notions of model would allow the deployment of set theory and model theory for the analysis of scientific models, as well for the analysis of theories through axiomatic treatment.

Although initial presentations of the semantic view of theories relied on this set-theoretical predicate view, most current defenders of the view adopt what Lloyd calls the *state-space approach* (Lloyd, 1984). The standard presentation of the state-space approach asserts that the model is the state space associated with a set of dynamical equations (van Fraassen, 1980; Lloyd, 1984, 1994). Strictly speaking, what is being associated with the dynamical equations is the set of trajectories in the state space, or the *trajectory space*.

To fully understand this picture, we need to unpack all of the key terms beginning with the concept of *state*. Real, physical systems can be said to be in a particular state at a particular time. Intuitively, the state is a complete description of the properties of this system. We can call a system's *total state* all of its properties at a particular time. We are often interested in states associated with a particular scale and corresponding science, rather than a total state. So we may speak of a system's *thermodynamic state*, which would correspond to its temperature, pressure, entropy, and other macroscopic properties. Or we might speak of its *quantum mechanical state*, in which case we would be describing all of its microscopic, measurable properties entertained by quantum mechanics.

Physicists sometimes deal with the complete thermodynamic, or classical, or quantum mechanical state of a system. But in other sciences, it is almost always necessary to think about *scope-restricted states* of a system. For example, a biologist might be interested only in the population abundance of two species, even though the biological system has many other properties.

The notion of state has its home in real-world systems, but such states are mirrored in models. Models' states are sets of values of variable quantities. These are often intended to represent states of real systems, but this is additional and optional. More formally, the state of a model is a vector corresponding to determinate values for each determinable of the model system.

State spaces are the sets of all possible states of a system or a model. The independent dimensions along which a state can vary give rise to the dimensions of the state space. For example, the Lotka–Volterra model can independently vary along predator population abundance (P), prey population abundance (V), and time (t). Thus each state of the system corresponds to a point in the space. An examination of which points exist in the space tells the theorists which states are allowed by the model. Typically, a compact function such as a force law or governing equation can be used to compactly express allowed states as well as transitions between these states.

Since state-space dimensions correspond to the variable quantities in algebraic expressions, scientists often talk about these dimensions as corresponding to models' variables. Strictly speaking, variables are properties of model descriptions (see Section 3.2), but these correspond to determinable states of the model. So we really should say that a model's state is a determination of a set of determinables, but it is often more convenient to speak of models' variables. Such variables (read "determinables") correspond to quantities whose behavior the model accounts for. A model might tell us, for example, that a variable depends linearly on another variable. Model descriptions also contain parameters, which are fixed and taken to be exogenous, outside the scope of what the model can represent directly. For example, in the Lotka–Volterra model, the birth rate is a parameter because nothing in the model informatively describes what gives rise to the birth rate.

Finally, in the Lotka–Volterra model and most other examples discussed by philosophers of science, the state space has a temporal dimension. When this dimension exists, an evolution function can be constructed which takes time as an argument and maps states onto states. This function then describes the temporal evolution of the system. Curves constructed with this function can be described as the *trajectories* through the space.

Trajectories through the space describe how the model's states change through time. For example, one trajectory of a deterministic population model might begin from some point corresponding to the population's initial size, then travel through the space, with corresponding increases and decreases in

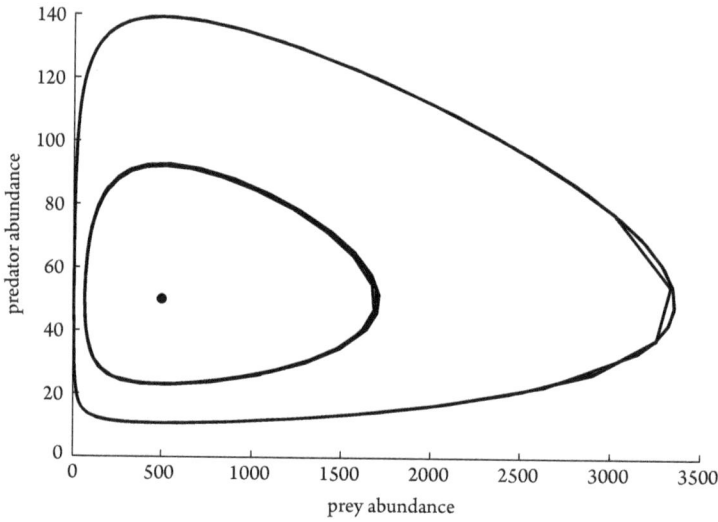

Figure 3.1 A representation of the phase space for the Lotka–Volterra model.

population size. The initial state, sometimes called the *initial conditions* of the system, will determine which path through the trajectory space that the model takes. A full trajectory space will correspond to all of the trajectories that could be generated from any set of independent variables and a fixed set of parameters. Different sets of parameters will give a different set of trajectories.

"Models as trajectory spaces" is thus the standard view in the literature about mathematical models. But is this really a sufficient characterization of the entirety of mathematical modeling? I think not, for several reasons.

First, trajectory spaces are only one kind of state space that is important in mathematical modeling. In some areas of modeling, the relevant structures do not have a proper time dimension. For example, for systems engaged in periodic motion such as springs, pendula, and the Lotka–Volterra model, it is often more useful to characterize the model in terms of the phase of the system. So the time dimension is dropped, and the state space becomes a phase space as shown in Figure 3.1.

Phase spaces still have an underlying temporal dimension, but some dynamical models use a dynamical dimension that isn't temporal. For example, in the study of chemical reaction dynamics, the relevant state spaces have dimensions corresponding to the relative positions of each atom in the reacting molecules and the energy corresponding to those configurations. These spaces are called *potential energy surfaces*. Paths through this state space correspond to particular reaction mechanisms, the intermolecular and intramolecular motions that occur during the course of a chemical reaction. Although a given set of molecules could move in one direction along a particular path, the state space itself is neutral as to the temporal dimension. The same set of molecules could move backwards along this path, or get stuck at a stable intermediate position.

Although the state spaces used in reaction dynamics are not temporal, they are still dynamical in character and can be thought of as paths in a state space. But why confine the notion of mathematical model to dynamical model? While we are often interested in how things change, sometimes we are interested in static, structural features of a system. In that case, other mathematical objects such as graphs and groups are more informative. Indeed, important and informative models of real systems can be made with these objects.

One example of the use of mathematics in a nondynamical model comes in the application of graph theory in modeling. A topic of considerable interest in recent years has been the effect of differing connectivities on very general properties of systems, including their robustness to perturbations and the average number of links between any two nodes in randomly constructed graphs. These mathematical objects are intended to stand for social networks, disease vectors, ecosystems, the internet, and so forth (Watts & Strogatz, 1998). They are clearly mathematical models, but their properties of interest are static attributes, not dynamical properties.

In sum, we can say that mathematical models contain mathematical structures which can be used to represent states and relations between states, especially transitions. In many cases of modeling, these mathematical structures are trajectories in state spaces, but other kinds of mathematical structures can be employed as well.

Before turning to the next kind of model, I want to say a few words about the ontological commitments of my view. This is not a book about the ontology of mathematics and I would prefer to remain as neutral as possible about such matters. For my purposes, I need only to assume that it makes sense to talk about mathematical objects such as functions and state spaces and that these objects are specified by the equations we use to describe them. Any metaphysical theory, even a highly deflationary one, that allows us to make sense of these aspects of mathematical practice is sufficient for my purposes. Which of these theories is true, of couse, is a much more complicated issue.

3.1.3 Computational Structures

A third kind of structure that can form the basis of a model is a computational structure.[1] In this book, the relevant kind of computational structure is an *algorithm*, which is a set of instructions for carrying out a procedure. Computational

---

1. The term 'structure' has a formal meaning in the theory of computation, but that is not the way I will use it here.

models are thus sets of procedures that take a starting state as an input and specify how this state changes, ultimately yielding an output.

Stated in these terms, it looks like there is little difference between mathematical and computational models. At an abstract level, this is almost certainly true and I am perfectly happy to say that computational models are a subset of mathematical models, albeit an especially important subset. However, what is distinctive about computational models is that the procedure itself is the core component of the model, the structure in virtue of which parts of a target can be explained (Kimbrough, 2003).

Schelling's model of segregation nicely illustrates the point. Say we find a pattern of segregation in a city that looks much like what Schelling predicted. If his model is an explanation of that pattern of segregation, then the model's algorithm must be similar to what is happening in the city. Model agents follow a two-step procedure: (1) determine whether they have enough similar neighbors to satisfy their utility function, and (2) if their function is unsatisfied, move to a new position. The model will explain the segregation in a real city if an analogue to that procedure characterizes the real agents in that city.

While mathematical structures could play this procedural role in principle, there are several features of computational structures that make them especially suited to the task. One such feature is their conditional structure. For example, in the Schelling model of segregation, agents assess whether or not their utility function has been maximized. If it has, they remain in the same state. If it has not been maximized, then they move to a new position, creating a new state of the model. This kind of conditionality is very naturally represented procedurally, but difficult to represent in many kinds of commonly used mathematical structures, including the functions represented by differential equations.

Another key feature that can be included in computational structures is probabilistic transitions. There are two reasons for including probabilistic transitions in computational models. One reason is that the modelers' intended target actually behaves probabilistically. More commonly, the intended target is actually deterministic, but too complex to be represented deterministically because of informational or computational limitations. In such cases, modelers can deploy probabilistic transition rules, and then study distributions of outputs starting from sets of randomly generated values.

For example, in our computational analogue of the Lotka–Volterra model, Ken Reisman and I represented individual organisms rather than populations. But this meant that we had to say something about whether a particular individual would be born, die, eat, or be eaten for each cycle of the model. So we assigned probabilities to each of these possibilities for each organism at each time step. In the aggregate, these transitions corresponded to the Lotka–Volterra model's treatment of these properties as deterministic, but population-level (for more details, see Section 9.3.3).

One final feature of computational structures that can be used to explain targets is their ability to represent parallel processes, where multiple elements of the process are carried out concurrently. Social processes of many sorts take place concurrently, with individuals making simultaneous, independent decisions based on the information that they currently have available. This can be modeled with a parallel computational structure, yielding true concurrency, or approximated using randomized serial processing.

Once again, Schelling's model of segregation is an excellent example. In his original model, agent movement followed a serial pattern in each round, starting from the top-left agent, and then moving across and down. Contemporary implementations of his model on computers typically use approximations to concurrency, by randomizing the order in which agents move in each round. However, if his model were run on a parallel computer, the process could be run with true concurrency (Tanenbaum & Van Steen, 2002).

I have talked about some of the features of computational structures that allow them to serve as models. As I said at the beginning of this section, computations can be described mathematically; indeed the standard way to talk about computation in the abstract is to talk about a set of states and transitions between these states. However, what I want to flag by distinguishing computational models from mathematical models (or treating them as a subset if you prefer) is that in computational models, the procedure is the core structure.

I want to close this section by reiterating that computation doesn't necessarily imply being implemented on a computer. Although most computational models are studied with computers, many computational models, such as Schelling's, were studied before the widespread use of computers. Moreover, most contemporary mathematical models are also studied using computers, because computers are excellent at doing mathematical calculations.

In summary, computational structures at the heart of computational models are procedural. They represent causal properties of their targets by relating these causes to procedures. They can be conditional, probabilistic, and parallel, all of which are difficult to represent using nonprocedural structures.

3.2 ■ MODEL DESCRIPTIONS

The second component of model anatomy I will discuss is the *model description*, the equations, pictures, or words that describe a model. As an example, consider the concrete San Francisco Bay model. The model can be represented by blueprints, technical drawings (Figure 3.2), and even pictures (Figure 3.3). These are some of the model descriptions of the Bay model. The Lotka–Volterra model is an interpreted mathematical object, but it is described by Equations 2.3 and 2.4. Schelling's model of segregation is a set of procedures,

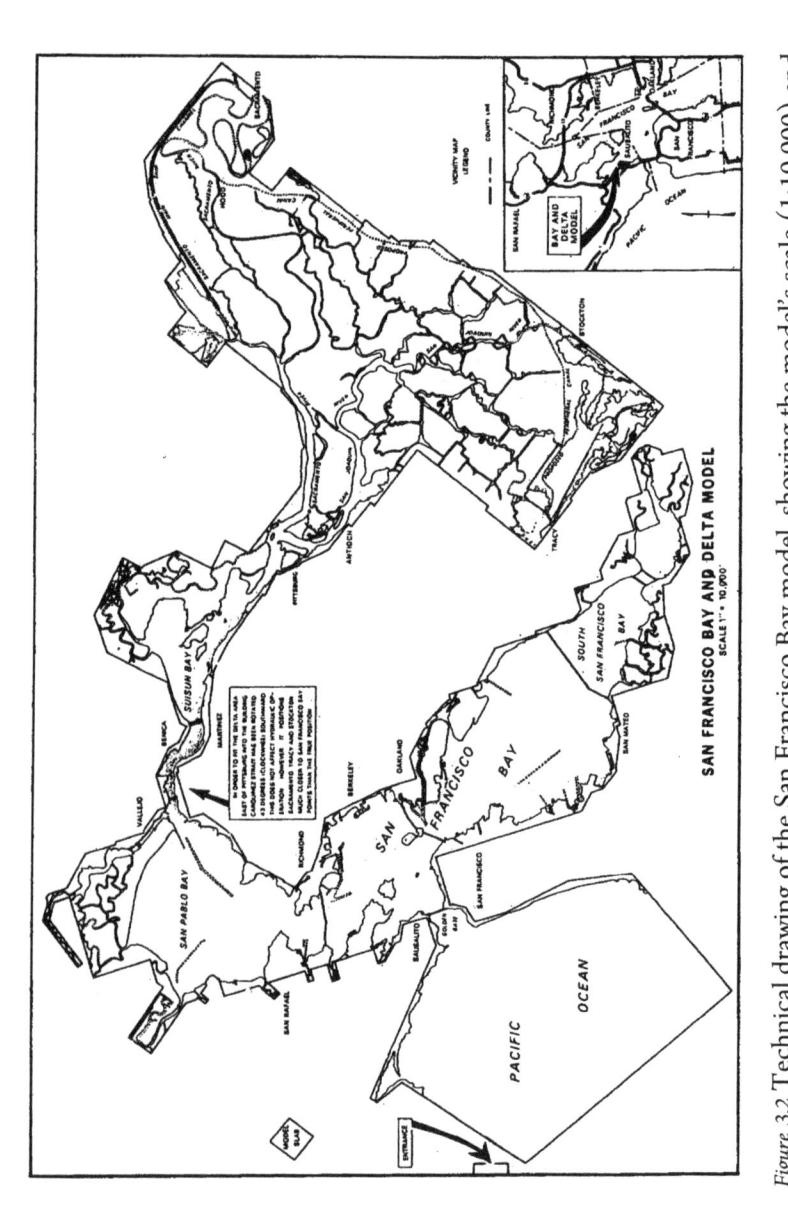

Figure 3.2 Technical drawing of the San Francisco Bay model, showing the model's scale (1:10,000) and orientation. The portion of the model representing the Suisun Bay and San Joaquin Delta was rotated 43 degrees so that it could fit in the warehouse. No. 91-012; "Bay Point Erosion and San Francisco Bay Study Project Files, 1946–1965;" Civil Works Project Files, San Francisco Bay Study, National Archives and Records Administration–Pacific Region (San Francisco).

Figure 3.3 A segment of the San Francisco Bay model, showing its representation of the Golden Gate Bridge.

but these are described by a program's source code or a set of statements. When we talk about models, write about them, or show a picture or diagram, we are employing a model description. These descriptions must be distinguished from the models themselves.

One of the first philosophers to emphasize the importance of model descriptions was Ronald Giere. In his account, model descriptions are referred to as the set of statements that *define* the model (Giere, 1988, 83). His examples are mostly mathematical, and he takes model descriptions to be an equation or set of equations in these cases. By asserting that the relationship is one of definition, Giere seems to be implying two things. First, the relation between models and model descriptions is stipulative: Models are created by their descriptions. Second, this relationship involves logical satisfaction: Every part of the model has the properties and relations asserted of it by the model description.

Giere's account was developed with mathematical models in mind, but it can be extended with modification to concrete models. The stipulation part of his account cannot really be preserved because concrete models are physical objects and exist independently of their descriptions. However, the satisfaction part of the account can be preserved. On this view, concrete models would have to satisfy their descriptions by containing the properties and relations asserted by their descriptions. Alternatively, if we take the model descriptions to be interpreted, we could say that the model descriptions truthfully describe properties of models.

This is a reasonable place to start constructing an account of models, but is ultimately incomplete, because in almost every case, models have properties not mentioned in their descriptions. For example, the Bay model is likely to have many small imperfections in its concrete or its steel frame that are not mentioned in its description. Giere's account is committed to saying that in such a case, the diagram is not a model description of the actual model. But this awkward conclusion seems at odds with the behavior of the scientific community, which has conventions about how to read technical diagrams with differing levels of tolerance for mismatches between descriptions and models. It seems likely that scientists would conclude that such a diagram isn't inaccurate, it is simply silent about these details. In this respect, the diagram is more abstract then the model. It speaks truthfully about some aspects of the model, but is silent about others.

Thus we must amend Giere's account in order to deal with the fact that model descriptions can be more abstract than models. We need to know what is abstracted or not being mentioned on the one hand, and what is being asserted to be absent. For example, if the diagram of the Bay model does not include a drawing of the water pump at the mouth of the Bay model, it is not clear whether the diagram is being noncommittal about the existence of the pump or asserting that it should not exist.

I think the best way to deal with these ambiguities is to accept from the outset that a model description alone isn't sufficient to pick out a model or set of models. Instead, a combination of the model description and a theorist's intentions about how the description is to be interpreted specify the model (see Section 3.3 for more details about the contents of the interpretation).

Another place where Giere's account requires some modification is in the nature of the mapping between model descriptions and models. Since Giere regards model descriptions as definitions of models, the relationship must be bijective. But this seems too strict because in general, there is neither a one-to-one correspondence nor even a requirement of surjection between model descriptions and models. A single model can be described in many different ways, such as with blueprints, pictures, equations, or computer programs. For example, the San Francisco Bay model was initially represented with blueprints, but in published discussions of the model, most of its features were described in text. As the Army Corps has moved from using physical models to a greater reliance on computational models, their model descriptions were given in computer code. Thus many different model descriptions can describe a single model.

Just as a single model can be associated with multiple descriptions, a single description can and usually is associated with multiple, distinct models. Consider again the complete set of blueprints for the Bay model. Even if these blueprints and the accompanying commentaries are very detailed, they are likely to be realized equally well by multiple models. A single model description can pick out multiple, nonequivalent models if there is any vagueness or imprecision in the model description. These uncertainties can be small or large, but they affect how many models the description is associated with. A model description of maximal precision describes a single model perfectly, and any amount of imprecision in the description allows a single description to pick out multiple models. In general, the less precise or less specific a model description is, the greater number of models it is associated with. Thus the relationship of model descriptions to models is many-to-many.

Given that the relationship between model descriptions and models is many-to-many, what is the best way to think about the relationship between models and model descriptions? At its base, the relationship is representational; model descriptions represent models. And while the relationship is not one-to-one between models and descriptions, there is still a very tight link. Thus I will speak of model descriptions as *specifying* models, and of models as *realizing* model descriptions. As model descriptions and construals are refined and made more precise, they specify models more finely.

To illustrate the specification relation more fully, we can consider the San Francisco Bay model's description. Since a good documentary record exists, we can see exactly what the Army Corps of Engineers believed was necessary to specify the model, both in terms of the actual model description and its accompanying commentary. The best record of their model description is contained in the *San Francisco Bay–Delta Tidal Hydraulic Model: User's Manual* (Army Corps of Engineers, 1981). In addition to photographs and technical drawings, a section titled "Description of the Bay–Delta Model" contains the following information:

> The Bay–Delta Model occupies an area of about one acre, and is completely enclosed in a 128,500 sq. ft. shelter [...]. The limits of the model ... encompass the portion of the Pacific Ocean extending 17 miles west of the Golden Gate, San Francisco Bay, San Pablo Bay, Suisun Bay, and all of the Sacramento–San Joaquin Delta east of Suisun Bay to the cities of Sacramento on the north and east, Stockton on the east, and Tracy on the south.
>
> The model is approximately 320 feet long in the north-south direction, and about 400 feet long in the east-west direction. All important features of the San Francisco Bay and Sacramento–San Joaquin Delta are reproduced including all ship channels, rivers, creeks, sloughs, and canals in the Delta, and all major wharves, piers, slips, dikes, bridges, and breakwaters located throughout the system. The Sacramento

TABLE 3.1. *Scale relationships for the San Francisco Bay model*

Model	Factor	Prototype
1 ft.	Depth	100 ft.
1 ft.	Length or Width	1,000 ft.
10	Slope	1
1 sq. ft.	Area (cross section)	100,000 sq. ft.
1 sq. ft.	Area (plan)	1,000,000 sq. ft.
1 sq. ft.	Area (plan)	23 acres (approx)
1 sq. ft.	Area (plan)	0.03 sq. mi. (approx)
1 cu. ft.	Volume	100,000,000 cu. ft.
1 cu. ft.	Volume	2.296 ac. ft. (approx)
1 cu. ft. per sec.	Discharge	1,000,000 cu. ft. per sec.
1 gal. per min.	Discharge	1,000,000 gal. per min.
1 ft. per sec.	Velocity	10 ft. per sec.
1	Salinity	1
1	Time	100
0.6 min.	Time	1 hr.
1 min.	Time	1 hr. 40 min.
14.4 min.	Time	1 solar day (24 hrs.)
14.9 min.	Time	1 lunar day (24 hrs. 50 min.)
14.9 min.	Time	1 tidal cycle

River to Verona ... and the San Joaquin River to Vernalis ... are also modeled. In addition, the inflows for the Mokelumne and Cosumnes Rivers, and export pumping facilities discharching into the Delta-Mendota Canal, the California Aqueduct, and the Contra Costa Canal are reproduced in the model (Army Corps of Engineers, 1981, 6–2).

The engineers go on to discuss the agriculture drainage flows which are included in the model, and then give a table describing the scale relationships, reproduced here as Table 3.1. Finally, the engineers explain the materials with which the model was built, and the appurtenances necessary to reproduce the tidal currents and salinity distribution. These are features which are necessary for the operation of the model, but whose presence is certainly not taken to exist in the target.

Obviously I have reproduced only a small part of the full model description given by the Army Corps. But even with the complete description given in the operator's manual, it would have been impossible to build a model without far more detail about how the model was actually constructed, the materials used for the model, the model's pumps, and so forth. Such a description could obviously be satisfied by more than just the specific model in Sausalito.

The forms of model descriptions are similar, although often less detailed, for mathematical and computational models. Equations or other kinds of statements specify mathematical objects and these objects satisfy their descriptions. However, unlike in the case of concrete models, mathematical models can be studied and manipulated only via their descriptions. While the Lotka–Volterra model itself is not a set of equations, it can be studied only through proxies such as these equations. This is probably the main reason that scientists often informally refer to equations as models; their attention is focused on these equations.

When dealing with dynamical models such as Lotka–Volterra, model descriptions of mathematical models can be written down with more precise and compact statements than those associated with concrete models. Ultimately, algebraic statements such as differential equations are functions that map values from the domain to the range. As I discussed in Section 3.1.2, this is typically thought of as the specification of trajectories in a state space. But an algebraic expression with no substitution of values for dummy letters can specify an infinite set of infinite sets of trajectories in a single state space. So we need to make a further refinement.

Steven Orzack and Elliott Sober introduced the helpful concept of model description *instantiation* (Orzack & Sober, 1993).[2] An uninstantiated model description is an equation in which values are not assigned to the parameters. Instantiating a model description means adding in values for the parameters. A fully instantiated version of the Lotka–Volterra model description has values set for each parameter. The population abundances (P and V) are the dependent variables of this description so they remain as algebraic symbols even when the description is instantiated. Two different instantiations of the Lotka–Volterra model are described by the following equations:

$$\frac{dV}{dt} = 0.01V - (1.0V)P \qquad \frac{dP}{dt} = 0.5(1.0V)P - 0.01P \qquad (3.1)$$

$$\frac{dV}{dt} = 0.1V - (1.0V)P \qquad \frac{dP}{dt} = 0.5(1.0V)P - 0.001P \qquad (3.2)$$

Instantiations of uninstantiated model descriptions give rise to hierarchies of models and their descriptions in the following way: Uninstantiated model descriptions specify families of instantiated model descriptions. Depending on the precision of these model descriptions, a single model or a family of models can be specified by a description. If the parameters are set with some imprecision, such that they can take a range of values, then the description will specify a subset of the models specified by the uninstantiated description. If we instantiate the parameters with completely precise values, then the instantiated description will specify a single model.

2. More precisely, Orzack and Sober described the distinction as one between uninstantiated and instantiated *models*. However, since they do not distinguish between model descriptions and models, I take it that my use of these terms is in the spirit of their discussion.

Model descriptions for the third type of model, the computational model, are also usually given abstractly. They often take the form of an explicit description of the model's basic procedure using a programming language's source code, pseudocode, or in simple cases, discrete mathematics. For example, the following is a few lines from the source code of the individual-based Lotka–Volterra model described in Section 9.3.3:

```
to prey-reproduce   ;; predator procedure
  ifelse foliage?
    [if random-float 100 < prey-conversion-prob [
      hatch 1 [ rt random-float 360 fd 1]]]
    [if random-float 100 < prey-reproduction-prob [
      hatch 1 [ rt random-float 360 fd 1]]]
end

to pred-reproduce   ;; prey procedure
  if random-float 100 < pred-conversion-prob [
    hatch 1 [ rt random-float 360 fd 1]]
end
```

Summing up, models are specified by model descriptions. The relationship of specification is many-to-many because a single model description may specify multiple models and an individual model may be specified by many kinds of descriptions. While there is considerable latitude for the kinds of representations that can serve as model descriptions, concrete models are usually represented with concrete model descriptions, and mathematical and computational models are usually represented with abstract (algebraic or pseudocode) descriptions.

Although it is natural to think of model descriptions as being set down before concrete models are constructed or found, this is not strictly necessary. In some cases, the model is constructed before or without a description. In others, the description comes first. And perhaps most commonly, the two are produced in tandem. When Watson and Crick developed their model of DNA, they constructed the model before its description. In fact, the key to solving the structure of DNA involved seeing physical characteristics of the model and using these to think about the ways that the backbone and nucleic acids of DNA could arrange themselves. Only after the physical model was constructed was a mathematical description of the model written down, in that case to check the validity of the structure from X-ray crystallography data.

In other cases, the description of the model preceded the construction of the model. For example, the Army Corps of Engineers constructed the Bay model by first making detailed technical drawings. The model was then constructed according to the specifications of these drawings. When the Reber plan and other saltwater barriers were studied, modifications were made to the model. In these cases, the Corps worked from technical drawings about the temporary modifications they would make in order to evaluate the soundness of these saltwater barriers. So there is no general order for the construction of models and their descriptions; differing circumstances dictate creating one or the other first. And in some cases of concrete models, descriptions need not be generated at all.

We have now seen that the first major component of scientific models is structure. There are three kinds of structures and each one of these kinds of structures can be specified using abstract or concrete model descriptions. The relation between model descriptions and models is many-to-many, so model descriptions only partially specify models. I will next turn to modeler's intentions, which are needed to fully specify models.

3.3 ■ CONSTRUAL

As I said at the beginning of this chapter, models are a combination of structure and interpretation. I have already introduced the three kinds of structures that modelers employ. In this section, I will discuss modelers' interpretations, what I will call their *construals*. Construals provide an interpretation for the model's structure, they set up relations of denotation between the model and real-world targets, and they give criteria for evaluating the goodness of fit between a model and a target. I take up these aspects in turn.

Construals are composed of four parts: an *assignment*, the modeler's intended *scope*, and two kinds of *fidelity criteria*. The assignment and scope determine and help us to evaluate the relationship between parts of the model and parts of real-world phenomena. The fidelity criteria are the standards theorists use to evaluate a model's ability to represent phenomena.

Assignments are explicit specifications of how parts of real or imagined target systems are to be mapped onto parts of the model. This explicit coordination is important for two reasons. First, although the parts of some models seem naturally to coordinate with parts of real-world phenomena, this is often not the case. For example, harmonic oscillator models were first developed to make predictions about the periodic motion of physical systems, but as mathematical models, they remain abstract objects without obvious analogues to the properties of springs, molecules, or even pendulums. Further, chemists use harmonic oscillators to model vibrations in bonds. These models represent atomic positions as points in a coordinate system and describe the periodic offset of these points, which corresponds to molecular vibration, as a harmonic oscillator. The assignment is a formal record of this type of coordination.

$$\frac{dV}{dT} = rV - (aV)P$$

- V: prey population density
- $(aV)P$: functional response

$$\frac{dP}{dT} = b(aV)P - mP$$

- P: predator population density
- $b(aV)P$: numerical response
- mP: predator death rate

Figure 3.4 Model description for a family of Lotka–Volterra models with the assignment made explicit.

Assignments are often not made explicit in discussions of models. This is because communities of modelers have standard conventions for reading model descriptions. Where conventions are not explicit, where they are being violated, or where the modeler needs to be especially specific, she will be forced to make the assignment explicit in discussions about the model, such as is shown in Figure 3.4.

Models typically have structure not present in the real-world phenomena they are being used to study. Consider the Lotka–Volterra model of predation. This model's main dependent state variables are population abundances, but these states and the transitions between them are represented with continuous mathematics. This means that the model can describe transitions between states where the state variable is an irrational number. Volterra certainly did not intend to represent any real or possible population of fish with negative or irrational densities. Therefore in construing his model, Volterra only accepted rational values for the state variables (and probably only certain ranges of those numbers) to population abundances in the Adriatic and other possible populations.

Modelers thus make decisions about which aspects of their models are to be taken seriously. Their intended scope specifies which aspects of potential target phenomena are intended to be represented by the model.[3]

---

3. A similar position is taken by Suppe (1977a).

I can further illustrate intended scope by returning to Volterra's model. The model itself describes only the size of the predator population and of the prey population, the natural birth and death rates for these species, the prey capture rate, and the number of prey captures required to produce the birth of a predator. It contains no information about spatial relations, density dependence, climate, microclimate, or interactions with other species. If the scope is such that those features were intended to be represented, Volterra's model does a poor job, because it would indicate that there is no density dependence, no relevant spatial structure, and so on. By choosing a very restrictive intended scope and hence a narrow target, we indicate that Volterra's model is not intended to represent these features.

The third and fourth aspects of a model's construal are its fidelity criteria. While the assignment and scope describe how the real-world phenomenon is intended to be represented with the model, fidelity criteria describe how similar the model must be to the world in order to be considered an adequate representation. There are two types of fidelity criteria: *dynamical fidelity criteria* and *representational fidelity criteria*.

Dynamical fidelity criteria tell us how close the output of the model must be to the output of the real world phenomenon. By output, I simply mean the values of dependent variables in the model and in the world. Dynamical fidelity criteria are often specified as error tolerances. For example, a dynamical fidelity criterion for a predator–prey model might state that the population abundance of the predators and prey in the model must be $\pm 10\%$ of the actual values before we should accept the model. These criteria deal only with the output of the model, that is, its predictions about how a real-world phenomenon will behave.

Representational fidelity criteria are more complex and give us standards for evaluating how well the structure of the model maps onto the target system of interest. Typically, these criteria specify how closely the model's internal structure must match the causal structure of the real-world phenomenon to be considered an adequate representation.

For example, a biologist studying predation might want her model to capture the factors that would cause a predator to either initiate or terminate predation (hunger, food storage, satiation, etc.). If this were an important part of her representational fidelity criteria, the Lotka–Volterra would score very poorly because it says nothing at all about these factors.

A similar set of distinctions is made by engineers who build scale models. *Kinematic similarity* means that the rates of change of variables (often flows of some sort) are similar between model and target. *Geometric similarity* means that a model and its target are similar in structure. These two roughly correspond to my dynamical and representational fidelity. And finally, for engineers, *dynamical similarity* means that the relevant dimensionless quantities, such as the Reynolds or Cauchy number, are the same in model and target (Kline, 1986; also see Sterrett, 2005).

Along with a concrete, mathematical, or computational structure, theorists' construals generate models. To say that a model is structure plus interpretation means that models are structures whose parts are interpreted via the assignment. They can potentially denote parts of a target as specified by the theorists' intended scope, and they are evaluated by the theorists' fidelity criteria. These four components of the construal constitute the theorists' interpretation of the model.

3.4 ■ REPRESENTATIONAL CAPACITY OF STRUCTURES

So far in this chapter I have given an account of scientific models that involves two main components: structure and interpretation. By interpreting a concrete, mathematical, or computational structure, a theorist can investigate phenomena. A question arises: If interpretation drives so much of what makes something a model, can any structure at all be a model? And can anything be a model of anything else?

To a first approximation, the answer to these questions is "yes." Even the simplest objects or phenomena stand in many kinds of resemblance relations to other things. These resemblance relations can form the basis of the use of these objects as models, where properties of one thing stand for at least some of the properties of the other. Simple and complex machines, paper and plastic shapes, organisms, and highly complex scale models such as the Bay model all count as concrete models and all stand in myriad resemblance relations to real-world target systems. Similarly, an almost endless array of mathematical and computational structures can be used to represent different targets, both real and imagined.

Nevertheless, it is misleading to say that anything can be a model or that anything can represent anything else. For example, imagine a single colored marble sitting alone on a table. Now say that we were interested in using this marble as a model of the San Francisco Bay. We would quickly run into trouble. Unless one had impossibly low standards of fidelity, the shape of the marble cannot represent the shape of the Bay, which isn't even slightly spherical. Moreover, the dynamical aspects of the Bay such as water currents, changes in salinity and temperature, and depth changes due to shoaling cannot be represented by something static that has one state only.

Now consider the marble once again and a very different kind of system—a highly abstract mathematical one like the trajectories of the Lotka–Volterra model. There are some relations that hold between the marble and this trajectory space: the marble shares its topology with a point. But little else about the myriad mathematical properties of these spaces can be captured by a model.

Can the marble be a model of any system at all? Certainly it can. For example, a highly abstract target of light reflecting off of a nearly spherical surface could be modeled with an illuminated marble. A marble might also represent reflectance of other, non-spherical shapes. But its capacity to represent systems very much unlike itself seems limited in two ways: its structure is very simple and it doesn't have multiple states.

So while strictly speaking it might be true that anything can be a model of anything else, we should draw two lessons from the example of the marble. The first is that concrete objects with very simple structures and few states have a low *representational capacity*. That is, they are not able to represent very many systems, especially ones of a very different type. The other lesson is that, with concrete models, scientists must pay attention to the overall similarity of the model to the system being studied. It is very difficult to represent one system with a model very much unlike it.

We can draw similar lessons for mathematical and computational models, but the situation is somewhat different. Mathematical and computational models are limited by their representational capacities, but to a considerably lesser extent. One of the reasons that mathematics is such a powerful tool for science is its flexibility. The very same mathematical structure can be used to represent a pendulum, a spring, a vibrating molecule, a laser, the movement of water, and so forth. So the transition from a concrete model to a mathematical model will almost certainly result in a qualitative increase in the representational capacity of the model.

But again, we should be cautious in saying that any mathematical or computational model can represent any target system at all. One issue concerns how much structure the model needs to have in order to make predictions up to the standards specified by the fidelity criteria. To take a trivial case, imagine that a biological population first grows exponentially, and then levels off to a stable population size, corresponding to the carrying capacity of the environment. Such a scenario simply cannot be modeled with any reasonable standard of fidelity using a model whose mathematical structure is a pure exponential function, or indeed with any function that takes only one variable. So there is a minimum number of variables and a set of needed functional forms required to model this type of growth with any reasonable degree of fidelity. This type of representational capacity has been called *dynamical sufficiency* by Lewontin (1974) and Godfrey-Smith (Godfrey-Smith & Lewontin, 1993), and the *sufficient parameter* by Levins (1966).

In addition to the dynamical sufficiency of a model, which is really a matter of being able to represent changes of state sequentially, there is also the question of what types of causal structures can be represented. While sometimes theorists want only to make accurate predictions, at other times they are interested in knowing if a model can make the right predictions in virtue of capturing the target's causal structure. In other cases, they are concerned to represent underlying mechanisms of the target without ever hoping to have a reasonable ability to predict its behavior. This suggests that a second kind of representational capacity should be assessed: *mechanistic adequacy*, the resources of the model for representing underlying causal structures.

For example, in the Lotka–Volterra model, no individual organisms are represented, only population abundances. This means that certain potentially relevant aspects of a target system such as spatial structure, heterogeneous distribution of individuals, and so forth simply cannot be represented in the model. The model doesn't have enough variables to capture this structure. This idea generalizes beyond mathematical models to computational and concrete structures as well.

There are other aspects of representational capacity that are relevant for mathematical and computational models. Some kinds of mathematical structures are continuous, which means that infinitesimal changes in the domain are mapped to infinitesimal changes in the range. Many traditional mathematical models such as the ones found in classical mechanics and classical populations dynamics (e.g., the Lotka–Volterra model) use continuous mathematics.

However, many contemporary mathematical and computational models are based on discontinuous functions. The reason for this is threefold: First, in cases where differential equations describing a mathematical model cannot be solved, numerical approximation is required. When such an approximation is conducted on a computer, a discrete approximation of the differential equation must be used. Second, all computational models are discrete, because the transitions of state described by algorithms are necessarily discrete. Finally, and most importantly, many of the phenomena of interest to scientists are not continuous. There are not a continuous number of predators in a predator–prey system; whole organisms are described by integers. So, many contemporary models are based on discrete mathematics. The intrinsic representational capacity of discrete mathematics for describing discrete systems seems greater, even if continuous mathematics is more general.

Another dimension of differences in representational capacity involves determinism. Transition functions can be described either deterministically or probabilistically. This corresponds to the transition happening necessarily or with some probability. In the Lotka–Volterra model, increases of prey populations of a certain magnitude are necessarily associated with increases of the predator population, and then decreases in the prey follow. But this process could also be represented probabilistically, by representing individual organisms and their probability of having offspring, being eaten, and so forth. Representing deterministic processes probabilistically is especially valuable in cases where the value of some key variables is not known, when a phenomenon is aggregated, or when structural symmetries in the system are best handled probabilistically (Strevens, 1998). However, deterministic structures cannot be used to model probabilistic processes unless the probabilities somehow cancel each other out when one is dealing at the aggregate level. Thus there is an asymmetry in the representational capacity of probabilistic and deterministic structures.

So we can see that there are many considerations that go into the representational capacity of different structures. The art of good modeling involves not only choosing an appropriate construal for a model, but also choosing structures with representational capacities appropriate to the target or the theoretical task at hand.

In this chapter, I have argued that we should think of models as being composed of structure plus interpretation. Models are specified by model descriptions, which can take the form of words, pictures, equations, diagrams, or computer programs and are accompanied by legends. These model descriptions specify concrete, mathematical, or computational structures. Theorists' construals provide an interpretation of these structures that sets up relations of denotation to potential target systems (assignment), helps identify the particular aspects of the target on which the theorist focuses (scope), and sets the standards of evaluation (fidelity criteria).

4 Fictions and Folk Ontology

The Lotka–Volterra model is abstract and mathematical, not concrete and physical like the Bay model or Watson's and Crick's model of DNA. Exactly what kind of thing a mathematical model is, however, has been the subject of considerable debate in the literature about models. In Chapter 3, I developed a version of what I will call the *maths* view of mathematical models.[1] On this view, mathematical and computational models are interpreted, mathematical structures. However, in the last few years, a concrete or *fictions* view of mathematical models has become especially popular. Philosophers such as Roman Frigg, Peter Godfrey-Smith, and Arnon Levy have argued that the best way to understand mathematical models is as if they were more closely related to literary fictions than to bits of mathematics. I should note at the outset that this claim is not the same as the assertion that models describe fictional scenarios because they are idealized (Vaihinger, 1911; Suárez, 2009). Proponents of both the maths and the fictions views can make this latter assertion. Rather, the difference between the accounts is that proponents of fictions accounts think that mathematical modeling involves engaging with fictions in a way that is analogous to reading stories or watching movies.

In this chapter, I will begin by discussing the motivations and arguments for fictions accounts. I will then raise a number of objections against these views, and go on to consider how the maths account I developed in Chapter 3 can be modified to take into account some of the important insights raised by proponents of these accounts.

4.1 ■ AGAINST MATHS: INDIVIDUATION, CAUSES, AND FACE-VALUE PRACTICE

Proponents of fictions accounts can raise a number of objections against maths accounts of mathematical models. The first of these objections has to do with

---

[1]. In a recent review of this debate, Mary Morgan and Margaret Morrison (1999) introduce slightly different terms to divide accounts of mathematical models into two traditions. Proponents of *concrete* accounts of mathematical models take mathematical models to be something like imaginary structures that would be concrete if they were real (Hesse, 1966; Black, 1962; Campbell, 1957; Godfrey-Smith, 2006). What Morgan and Morrison call the *abstract* tradition includes accounts of models as set theoretical structures (Suppes, 1960a, 1960b), as well as those that take mathematical models to be trajectories through state space (van Fraassen, 1980; Lloyd, 1994).

model individuation. If models are simply mathematical objects, then when two distinct models use the same mathematics, we will not be able to individuate them as separate objects. This situation occurs frequently. Take, for example, the harmonic oscillator model. The same mathematics can be used both to describe an idealized spring and a chemical bond. It seems like proponents of maths accounts must conclude that there is a single model being applied to these two cases, because a single differential equation (model description) describes the same set of trajectories in a state space (the model) in both cases. However, common scientific usage would have us think that these models are similar but not exactly the same.

The second problem with the maths account is considerably more serious than the first and involves the way that causal information is encoded in models. Many traditional accounts of models in the semantic view demanded only that models be *empirically adequate*. In other words, they must be isomorphic to a mathematical representation of the empirical substructure of a real-world system. However, many scientists are realists and demand that unobservable state variables and causal structures be accurately represented by their models. It is one thing to have a model with which one can make accurate predictions; it is another to have a model that makes accurate predictions for the right reasons. Such a model would represent the real causal structure of the target phenomenon. Can this be done with purely mathematical objects?

The answer would seem to be no, at least when talking about the dynamical models central to most accounts of mathematical models.[2] Mathematical objects can have structural and relational properties, but not causal ones. They naturally show correlation, but not causal dependence. It is not obvious how they can distinguish between a properly formulated forward-looking causal path and a backward causal path, or even a common cause. So if models are mathematical objects, they will have difficulty with representing causal structure (Matthewson, 2012).

A final problem is accounting for the fact that theorists talk about their models in concrete terms. For example, when discussing a model of predation, a theorist will often describe two populations of organisms that have properties like birth rates and capture rates. These sound like biological properties of concrete objects, not mathematical properties of abstract objects. So a mathematical account of mathematical models will need to make sense of the concrete language that theorists uses then talking about abstract objects.

---

2. There are, of course, mathematical frameworks designed for modeling causal relations. Such frameworks require keeping track of conditional probabilities using directed graphs. This involves considerably more mathematical structure than the kinds of dynamical models which are the paradigms of the philosophical literature about modeling.

Indeed, much of the motivation for fictions views of mathematical models comes from attending to the ways that theorists talk and presumably think about their models. In a recent essay, Peter Godfrey-Smith (2006) argues:

> I take at face value the fact that modelers often take themselves to be describing imaginary biological populations, imaginary neural networks, or imaginary economies. An imaginary population is something that, if it was real, would be a concrete flesh-and-blood population, not a mathematical object (735).

Thomson-Jones (1997) has referred to this observation as the *face-value practice* of modeling. Modelers often speak about their work as if they were imagining systems and in so imagining learning something about the real world.[3] I think that all sides can accept this face value practice of modelers. We might even go further and suppose that modelers don't just speak about their models in this fashion, but often think about them as imagined systems that would be concrete if they were real. The face-value practice is thus a claim about scientific cognition, not just how theorists talk about models.

An especially clear example of this can be found in John Maynard Smith's evolutionary genetics textbook. In the course of describing a model of the accuracy of RNA replication, he reasons as follows:

> Imagine a population of replicating RNA molecules. There is some unique sequence, S, that produces copies at a rate R: all other sequences produce copies at a lower rate, r. (Maynard Smith, 1989, 22)

In these first steps, he asks us to think about a collection of RNA molecules undergoing the process of replication. Presumably, we can imagine them because we have had some prior experience with the properties of RNA. We also assume that whatever is standardly true (whether or not we know about it) of RNA is also true of this imaginary population. He then asks us to consider a restriction to our initial imagined population: The replication rate is not consistent through the population. Instead, it is sequence-dependent, and one sequence has a greater rate of replication than all of the others.

Maynard-Smith goes on to describe the model in greater detail:

> A sequence produces an exact copy of itself with probability Q. If x_0 and x_1 are the numbers of copies of S and non-S respectively, then ignoring deaths,
>
> $$dx_0/dt = RQx_0,$$
> $$dx_1/dt = rx_1 + R(1-Q)x_0$$

---

3. Sugden (2002) describes this slightly differently, arguing that modelers construct *credible worlds* and study the properties of these constructed worlds. See Knuuttila (2009a, 2009b) for further discussion and development of Sugden's position.

In writing down these equations, I have assumed that when an error occurs in the replication of a non-S sequence, it gives rise to another non-S sequence: that is, I have ignored the very small probability that a non-S sequence will give rise to an S sequence. (Maynard Smith, 1989, 22)

In this next step, Maynard Smith further constrains the model by giving more information about the nature of the replication of the RNA molecules. In particular, he specifies the probability of exact replication and, as a result, also specifies the probability of nonexact replication. To complete the model, he goes on to derive an equation that describes the preservation of optimal molecules in the population as a function of differing degrees of accuracy in replication.

Godfrey-Smith and other proponents of the fictions approach would argue that this is an especially explicit illustration of a typical way theorists go about modeling. First, Maynard Smith imagined the model that he was taking about, in this case a population of self-replicating RNA molecules. He then went on to mentally fill in specific properties of the model and, at the same time, wrote down equations which recorded these specifications. As he thought more about the model and analyzed it in detail, he was able to refine it and make it more specific, recording these refinements in the equations that describe the model. His mathematical analysis was ultimately the mathematical analysis of an imagined scenario.

The question then becomes: What is the best way to account for this way of speaking and thinking, as well as the uses to which models are put? Whatever account of the epistemology and ontology of mathematical models is correct, this kind of discussion and cognition must be accounted for. Thus I take the face-value practice to be a core issue around which discussions about the ontology of mathematical models can be framed.

In the next section, I will introduce the simplest version of the fictions account of mathematical models. I will argue that in order to assess this account, it needs to be enriched with some epistemological and metaphysical details and will discuss two recent attempts to do so.

4.2 ■ A SIMPLE FICTIONS ACCOUNT

It is easy to motivate a fictions account of models if we start from the face-value practice. The most straightforward fictions account says that mathematical models are imaginary systems that would be concrete if they were real. A biological model of population dynamics, on this view, despite being described using mathematics, is actually an imaginary population of organisms, much like a population in the real world. The Lotka–Volterra model of predation consists of an imaginary population of predator animals and prey animals. These imaginary populations have the properties explicitly attributed to them in the act of modeling—such as growth and death rates, numerical responses, and functional responses. All of their other properties are either inferred from what has been stipulated or constructed from the theorist's imagination.

This is very similar to the way we construct fictional worlds from a novel or other written text. The text contains only some of the details and the rest must be filled in by us in order to make a coherent story. J. R. R. Tolkien doesn't tell us whether Frodo is left- or right-handed, but he must be one or the other or ambidextrous. So in order to draw inferences from fiction, readers may have to fill in details, even if these details do not really matter to the story or the author (D. Lewis, 1978; Ryan, 1980; D. S. Weisberg, 2008). Similarly, in order to draw inferences from the model, the theorist mentally fills in additional properties. But often the theorist can simply leave these properties vague, as discovering these properties by analysis is part of the point of mathematical modeling. I will call this the *simple fictions* view.

Peter Godfrey-Smith has been one of the main proponents of this simple-fictions view in recent years. He argues that we should draw a direct inference from the face-value practice of scientists to the view that mathematical models are imagined systems.

> Although these imagined entities are puzzling, I suggest that at least much of the time they might be treated as similar to something that we are all familiar with, the imagined objects of literary fiction. Here I have in mind entities like Sherlock Holmes' London, and Tolkein's Middle Earth. These are imaginary things that we can, somehow, talk about in a fairly constrained and often communal way. On the view I am developing, the model systems of science often work similarly to these familiar fictions. The model systems of science will often be described in mathematical terms (we could do the same to Middle Earth), but they are not just mathematical objects. (Godfrey-Smith, 2006, 735)

There are several advantages of the simple-fictions account. The first is that it appears to solve one of the problems that seem to plague mathematical accounts: models are easily individuated. Each imaginary system is a model and such systems can be represented in many different ways using words, equations, picture, or graphs. Model descriptions will always underdetermine models conceived of in this way. But this poses no problem and may even be an advantage because imprecise model descriptions can be used to generate families of models with greater degrees of generality.

Another advantage of this kind of account is that the similarity relation between a model and the world is intuitive, just as in the concrete model case. A model is similar to a target phenomenon in the world just in case it resembles that target. It is not easy to give a formal analysis of this similarity relationship (see Chapter 8 for my attempt), but the basic idea behind it is the same as in the case of concrete models. On this view, although imaginary, mathematical models are physical systems that have structural and behavioral similarity relations to real-world targets. More elaborate accounts of the relation will point to the role of theorists' interpretations of different parts of the model's structure via their construals.

Finally, this account has the advantage of taking seriously the way that theorists refine model descriptions on the basis of the mental picture they have of model systems. Godfrey-Smith points to the writings of theorists who describe themselves as first thinking about the model, as if they have some kind of mental picture of it, and then proceeding to write down their model description (equations) on the basis of this mental picture, as in the example above from Maynard Smith's work. This is one of the most important insights of the fictions account.

The simple-fictions account of mathematical models thus has a good deal of appeal. It gives us an obvious place to begin an analysis of the model–world relation by analogy to physical concrete models and it helps make sense of a very common mode of discourse among modelers.

We can think of the simple-fictions account of mathematical modelings as an "epistemic" account of scientific practice, in the manner discussed in Chapter 2. Such accounts are philosophical reconstructions of a scientific activity that tries to remain true to practice, but that aims to explain why the practice is successful.

What the simple-fictions account doesn't give us is an account of the metaphysics of models. Godfrey-Smith has argued that this should remain an open question. Our account of mathematical models will have to be informed by the metaphysics of literary fictions. Models' fictional scenarios are no more or less worrisome than the imaginary objects in ordinary fiction. Metaphysicians and philosophers of language will ultimately need to provide an account of the metaphysics of all of these objects, and this account may well be deflationary. However, it is perfectly obvious that we can reason about these worlds, talk about them, analyze counterfactuals about them, and so forth.

But not all philosophers have wanted to leave this an open question. Many have concluded that at least some of the metaphysics needs to be articulated in order to explain how models can be compared to real-world targets and how inferences can be drawn from models. I am sympathetic to this call for enrichment because I believe that a central focus of our account of models should be a detailed analysis of how models can be compared to their targets. In the next section, I will discuss two ways of enriching the simple account of models as fictions.

4.3 ■ ENRICHING THE SIMPLE ACCOUNT

Developing the metaphysics and epistemology of fictions accounts can proceed in one of two major directions: *fictions as possibilities* or *fictions as products of scientists' imaginations.* There may also be a third option centered on Thomasson's (1999) recent work about the metaphysics of fiction. She argues that fictional characters are best thought of as abstract artifacts, not as metaphysical possibilities or the mental states of authors and readers. Imported into this debate, the view would be that simple fictions are best understood as abstract artifacts, created and sustained by scientists' mental states. Since this possibility hasn't yet been explored in the literature about models, I will not discuss it further, and instead will focus on the other two options.

The first of these two options relies on a David Lewis-style "fictions as possibilities" metaphysics. On this view, the cognitive activity of imagining a scenario is deemphasized and the account foregrounds the role of the possibilities themselves. According to this view, models are concrete, non-actual possibilities. One natural way to develop this account is to say that a mathematical model is a possible world or part of a possible world. Although Godfrey-Smith doesn't necessarily endorse this strategy, it is a very natural interpretation of his simple-fictions view. Although he occasionally talks about theorists' imaginations, most of Godfrey-Smith's discussion concerns the consequences of thinking of models as both concrete and not real. Possibilities can be concrete, but it is hard to understand how imaginings can be concrete.

The second option is to argue that by invoking fiction, theorists are making no metaphysical commitments beyond the existence of mental states whose contents are fictional. This option deemphasizes the ontological commitments of models as fictions and lets an analysis of modeling rest on theorists' imaginations. Mathematical modeling in this way puts little or no stock in possibilities and becomes more akin to storytelling and games of make believe. Such episodes are then understood as mental states, not as concrete things.[4]

The fictions-as-possibilities view does not have many proponents in the philosophical literature. A version of the view is developed by Contessa (2010), although he imposes some restrictions drawn from a view of models as abstract objects. However, several authors have endorsed the imagination view including Roman Frigg (2010), Arnon Levy (2012), and Adam Toon[5] (2010, 2012). In the next two sections, I will discuss two variants of the imagination view. Then, I will turn to my critiques.

---

4. Another option in the literature is to treat fictions as rules of inference (Suárez, 2009). I will not explicitly discuss this view because its proponents are primarily considered with giving an account of idealization that is compatible with some kind of realism. They take 'fictional' to be nearly equivalent to 'idealized.'

5. Toon distinguishes between cases where scientists are modeling a real system, and those where they are not. When scientists model real systems, they ask us to imagine things about these systems, rather than describing them. But when they are modeling nonreal systems, like in models of perpetual motion and three-sex biology, model-descriptions are like passages about fictional characters and the model is what he calls the "model world."

4.3.1 Waltonian Fictionalism

One way of the developing the idea that mathematical models should be tied to scientists' imaginations comes from applying ideas about literary fiction to mathematical modeling. Roman Frigg has argued that the conceptual apparatus of Kendall Walton's theory of fiction as a game of make-believe can be used to help us understand models as fictions and their relationship to the world.

Many of the same face-value issues that motivate Godfrey-Smith's account also motivate Frigg's. Like Godfrey-Smith, he points out that scientists often talk about models as if they were physical things. But he also draws on an older philosophical discussion (Campbell, 1957) that says that physical theory has a "physical character," which means that it cannot be understood without comprehending its physical instantiations. This physical character is not fully captured by mathematics. Hence, the view of mathematical models he defends regards models as "imagined physical systems, i.e., as hypothetical entities that, as a matter of fact, do not exist spatio-temporally but are nevertheless not purely mathematical or structural in that they would be physical things if they were real" (Frigg, 2010, 253).

So far, this sounds very much like the simple-fictions view developed employing the metaphysics of possibilities. But Frigg thinks that the metaphysical commitments of such a view are too substantial, so he seeks an alternative.[6] He finds this alternative in contemporary work about the philosophy of art, specifically Kendall Walton's pretense theory.

Walton proposes to deal with the hard metaphysical, epistemic, and linguistic questions about fiction by understanding fiction as akin to a game of make-believe. So when we want to evaluate the question "Is Mordor to the east of Gondor?" we engage in a make-believe scenario where Middle Earth is a place with spatial relations and geography. We use *The Lord of the Rings* and related books as *props* that license the "moves" we are allowed to make in this pretend game. Such authorized rules along with principles of generation allow us to evaluate the truth of such geographical claims. "Is Mordor to the east of Gondor?" means "In the game of Middle Earth make-believe, is Mordor to the east of Gondor?" This sentence is true assuming that the relevant community of Tolkein readers is engaging in the same game authorized by the text of *The Lord of the Rings*.

Frigg proposes to use the resources of this theory to understand mathematical models. Model descriptions serve as props for the relavant game of make-believe. In addition to the rules explicitly authorized by the model description, background theories and mathematics provide a further body of rules.

---

6. Frigg also gives other motivations for finding an alternative to the possible-worlds framework on pp. 256–257 of his (2010).

What is explicitly stated in a model description (that the model-planets are spherical, etc.) are the primary truths of the model, and what follows from them via laws or general principles are the implied truths; the principles of direct generation are the linguistic conventions that allow us to understand the relevant description, and the principles of indirect generation are the laws that are used to derive further results from the primary truths. (Frigg, 2010, 260–261)

The major advantage of Frigg's view over the simple-fictions account is that it give us a way of dealing with what look like worrisome metaphysical commitments of fictions; it seems to dispense with them entirely. Since games of make-believe are psychological, there is nothing extra to posit beyond humans' cognitive systems and their ability to use mathematics. This is in contrast to the Lewisian development of the simple-fictions view, where imaginary concrete systems are treated as possibilities.

While this does seem like an advantage to many philosophers of science who prefer a sparse metaphysics, it also seems to undercut the simple-fictions story that models are related to the world by physical resemblance. If mathematical models are games of make-believe, they don't resemble anything in the physical world because they are scientists' mental states. Thus, Frigg has to give us an account of how we can learn about real targets from games of make believe. This is a nontrivial matter because now we are owed an account of how something inside a modeler's head can be compared with the properties of a target.

The general problem of comparing real targets to imagined systems is what Frigg calls the problem of *transfictional propositions*. In our case, the specific issue is the problem of comparing the properties of imaginary systems to the properties of real-world objects. Frigg says that this ends up not really being a problem because we only have to

compare features of the model systems with features of the target system. For this reason, transfictional statements about models should be read as prefixed with a clause stating what the relevant respects of the comparison are, and this allows us to rephrase comparative sentences as comparisons between properties rather than objects, which makes the original puzzle go away. (Frigg, 2010, 263)

In other words, Frigg thinks that we can construct an abstract representation of the properties of both model and target. These properties are then compared to one another, rather than the model being directly compared to the target.[7]

Frigg thus gives us an elaboration of the simple-fictions account that seems to avoid metaphysical commitments beyond the psychological. It also promises to

---

7. As Godfrey-Smith notes (2009), this solution comes at a greater cost than Frigg acknowledges. Since these properties are properties of imaginary scenarios, they are themselves uninstantiated. It isn't clear that uninstantiated properties are on stronger metaphysical ground than uninstantiated objects and systems.

make sense of the face-value practice of theorists who regularly talk as if their models are fictions. But this account does not dispel worries about model–world comparisons and, as I will argue, it has a problem about variation between imaginations of theorists.

4.3.2 Fictions Without Models

A more radical Walton-style account of modeling has recently been proposed by Levy (2010). He argues that we can think of modeling as a kind of fictional activity, but one that doesn't really require the introduction of something new called a model. Instead, we can think of modeling as a special type of idealization.

He asks us to consider an example of Martin Nowak's recent work on the question of whether the architecture of macrocellular structures called crypts affect the probability of cancer. Nowak writes:

> One simple approach considers N cells in a linear array. At each time step a cell is chosen at random, but proportional to fitness. The cell is replaced by two daughter cells, and all cells to its right are shifted by one place to its right. The cell at the far right [dies]. The cell at the far left acts as a stem cell. (M. A. Nowak, 2006, 222)

Godfrey-Smith and Frigg would understand this passage as the introduction of an imaginary population of cells in an array, but Levy argues that there is another alternative. Nowak could be asking us to imagine real cells, understood in a specific way. He writes:

> But we might also take Nowak to be asking his readers to imagine a *real-world crypt*, that it is a linear array with the specified properties. Call this the *de re* reading of model descriptions. Rather than thinking in terms of two steps, specification and comparison, the de re reading treats the model description as directly about its empirical target. What we have here is much like imagining, of oneself, that one were prettier, or a world-class athlete. (Levy, 2012, 12)

On Levy's account, there really is no model at all. The practice of modeling is simply the practice of thinking about a target system in a special, nonveridical fashion. One needn't posit the construction of an imaginary target system nor even a mathematical structure. In this way, Levy's view is especially minimal in its philosophical commitments.

Levy motivates this account by arguing that the simple-fictions and Waltonian views, which he calls *de novo* views, do not have the resources to explain how models can be used to explain their targets. For example, Frigg's account requires appealing to properties shared between the fiction and the target. As I discussed in the last section, this is more complicated than it seems because, strictly speaking, fictions have no properties at all. Frigg must therefore appeal to uninstantiated properties in order to make the comparison, but this looks like the reintroduction of metaphysically problematic machinery. Levy argues that this puts Frigg's view on par with Lewisian views in that they both need to appeal to metaphysical properties beyond theorists' imaginations. Either we have to endorse something metaphysically robust and naturalistically suspect, or else we cannot actually make model–world comparisons.

In contrast, Levy argues that his own *de re* view does not have a bloated metaphysics and seems to be able to explain how modeling can tell us about their targets. There are no intermediate steps between theoretical representation and target, because theoretical statements are always directly about their targets.

4.4 ■ WHY I AM NOT A FICTIONALIST

Fictional accounts about mathematical models certainly have appealing properties. Both Frigg and Godfrey-Smith present views whereby it is easy to undersand how modeling can be a process of indirect representation and analysis. The analysis is indirect because there is a "stopover" in a model that is either constructed (Godfrey-Smith) or just imagined (Frigg). Moreover, at least on the simple-fictions view, we can fairly easily give an account of the model–world relation that is very similar to the kind of account we might give for concrete models: Models literally resemble their targets, hence what we learn about the model can be compared directly to the target. Levy's view doesn't have these features, but he does allow us to make sense of many modes of scientific reasoning where scientists invite us to think about a system as if it had certain properties. In this way, we can see models as functioning in the manner of metaphors, a position that has appealed to many philosophers over the years (e.g., Black, 1962, and Hesse, 1966).

Despite these very real attractions, I think fictions views suffer from some considerable problems that ultimately make them untenable. As I will indicate in my discussion, some of these seem surmountable, albeit with cost. The solutions to others, however, seem more problematic than simply accepting the maths account. There are four major problems that I will discuss: inter-scientist variation, the limited representational capacity of fictions, the inability of Levy's view to account for the practice of modeling, and variation in the face-value practice.

4.4.1 Variation

Insofar as models are fictions, there may be considerable differences in the way these are conceived of by different scientists. This has different consequences for the different elaborations of the fictions view. For the simple-fictions view with a Lewis-like metaphysics, it suggests that a model description will have to pick out an equivalence class of possible worlds and that theorists' imaginations will lock on to one or a small set of particular worlds. For Frigg, variations among the theorists will generate a set of differing games of make-believe. For Levy, there will be differences in the way that a target is being imagined *de re*.

Before the cases are dealt with individually, it is useful to note that when we are speaking about literary fiction, this kind of variation happens all the time and is often not a problem. If I think that Orcs have human-like feet and you think their feet look a bit more like bear paws, this doesn't pose a problem unless the shape of Orcs' feet becomes part of the story. Of course, if they did become part of the story, Tolkien would almost certainly have given us the necessary detail to understand how the story unfolded. Insofar as Tolkien and the story were silent on the issue, it remains an interesting thing to think about, the sort of thing people debate at fan conventions, but nothing critical turns on it. This possibility has led some people to distinguish between fictional worlds and possible worlds, arguing that fictional worlds simply are not determinant with respect to all properties (Eco, 1991).

Whether or not this kind of indeterminacy poses a serious problem depends on which facts are indeterminate. Discussions of literary cases have suggested the need to distinguish between *focal* and *nonfocal* properties of stories. The number of toes on Orcs' feet and the number of Lady Macbeth's children can remain indeterminate because they are not an important part of their respective stories. Nothing in the text turns on them and, at least as of yet, nothing that we might want to infer about the fictional worlds of Middle Earth or 16th-century Scotland depends on them either. So if my imagination differs from yours about these details, nothing is problematic. On the other hand, the fact that Rohan is to the west of Mordor plays an important role in *The Lord of the Rings*, even if this is never explicitly stated in the text. Disagreements between readers could not be tolerated about this or similar facts, or else important aspects of the story would be incomprehensible.

The equivalent issue for mathematical models concerns cross-scientist agreement about a model's focal properties. There has to be at least a high degree of consensus, if not complete agreement, about focal properties in order for modeling and model-based representation to work on the fictions view. So we might re-pose the problem of variation as asking whether there is significant variation in models' focal properties among scientists and if this prevents the necessary degree of agreement required for scientific inference.

Exactly how much of a problem this is depends on how fictionalists develop their accounts. One possibility is that proponents of fictions views might say that the mathematical description gives all and only the focal properties of the model. Everything else in the model is a free construction of the theorist's imagination. So while a high degree of variation in models is possible, all the variation would occur in the nonfocal properties.

This approach would address the problem of variation in a straightforward way. There would be no variation at all in focal properties, which presumably are the explanatory properties. Hence there is nothing special to explain about how theorists come to have consensus about the content of their models; all the important content is settled by the mathematics.

Although I agree that this is a possible solution, it comes at the expense of what is attractive about fictions accounts. Fictions accounts are motivated by the fact that the cognitive activity of modeling often seems to involve thinking about imaginary worlds which can be compared, in a relatively straightforward way, to the real world. As proponents of the fictions account like to emphasize, mathematical descriptions are extremely sparse. If the mathematical description exhausted the focal properties of the model, then models would be correspondingly sparse in their portrayal of fictional scenarios. Fictions then cease to look at all like real-world scenarios, militating against the claim that models can be compared to real systems in a straightforward way.

More plausibly, proponents of fictions accounts would claim that models have focal properties beyond what are in the model descriptions. Although not every property of the model-as-fiction is focal, models possess a wide range of focal properties. These properties, along with theorist-introduced nonfocal properties, are sufficient for constructing a fictional scenario rich enough to be compared to a real-world target.

If this is the case, then we need to ask what *principles of generation* get us from model descriptions to richer worlds populated with focal properties. Frigg suggests that the model description gives us the "primary truths" of the model, and that laws of nature fill in the rest. But this cannot be sufficient to generate the entirety of the relevant fictional scenario, even in simple cases. For example, laws of nature are not a sufficient supplement to the Lotka–Volterra equations to generate imagined populations of organisms. Such organisms need to have physical forms, metabolic and regulatory processes, behaviors, locations, and so on. No laws of nature can generate these properties simply from the Lotka–Volterra equations. So something beyond laws of nature is required.[8]

In philosophical literature about fiction, there are two standard accounts of how focal properties are generated. The first employs the *reality principle*, which says that the fictional world is filled in with everything from our own world unless the text specifically tells us to deviate from our world (Walton, 1990; see also discussion in D. S. Weisberg & Goodstein, 2009). This is a controversial move in the literary-fiction debate and is not widely accepted, but it seems even less plausible in the science case. Returning to the Lotka–Volterra model,

---

8. In conversation, Frigg has said he has a very liberal conception of laws here, and means to include background concepts and theory. I am still uncertain how this could work, although it might involve an indirect appeal to the reality principle or the mutual-belief principle.

even if we adopted the reality principle, how would this get us to an imaginary population of predators and prey? What properties would they have beyond what the mathematics gives us? How do we know to ignore some aspects of the mathematics, such as noninteger values of population sizes? These are all essential parts of understanding the Lotka–Volterra model and how it is used to explain biological phenomena, but cannot be generated from the real world plus the mathematics of the model. Indeed, some of these run counter to the mathematics.

The alternative view, introduced by Lewis (1978) and elaborated by Walton (1990), says that the construction of focal properties requires the *mutual-belief principle*. This principle says that:

> [i]f $p_1,\ldots,p_n$ are the propositions whose fictionality a representation generates directly, another proposition, q, is fictional in it if and only if it is mutually believed in the artist's society that were it the case that $p_1,\ldots,p_n$ it would be the case that q. (Walton, 1990, 151)

The idea here is that we fill in details in the fictional world not by importing all of our own world, but by importing what would be believed in the artist's world. The analogue in a scientific case would be something like the following:

> If $p_1,\ldots,p_n$ are the propositions that are generated directly by a model description M, another proposition, q, is warranted in that model's fictional scenario if and only if it is mutually believed in the scientists' community that were it the case that $p_1,\ldots,p_n$ it would be the case that q.

Undoubtedly there are better ways to formulate this principle, but it is clear that something along these lines will generate a fictional scenario that is as determinant as the intersection of the community's beliefs about what is focal to the model beyond what is in the model description. This will certainly limit variation, although it is unclear how much of a solution this really is. To test it, I will return to the example of the Lotka–Volterra model.

Let's consider two specific focal properties that are neither mentioned in the model description of the Lotka–Volterra model nor are they direct logical consequences of the mathematics. The first is the property that biological populations have integer numbers of members. This property is absolutely essential to interpreting the mathematics of the Lotka–Volterra model, yet it doesn't come from the mathematics. If one solves the differential equations at the heart of the model description, one gets predictions of noninteger numbers of organisms at some time intervals (see Section 3.3). The discrepancy between the mathematics and an integer-only interpretation of the model is implicitly accepted by the scientific community. Thus one way to understand the mutual-belief principle discussed above is as a codification of the implicit part of theorists' construals of their models. Ecologists mutually believe that the

Lotka–Volterra model should be interpreted as only giving predictions about integer numbers of organisms.

Now consider a different kind of focal property: the spatial structure of the two species in the predator–prey system. The mathematics of the Lotka–Volterra model say nothing at all about spatial structure, so on the maths view, the model is simply abstract with respect to this structure. But on a concrete view, the model is composed of real organisms and real organisms have to be placed in space.

Determinate spatial positions are necessary on the concrete view, but are they focal? I believe that they must be. Subsequent research on predator–prey models has shown that when spatial structure is included in the mathematics, only certain kinds of spatial structures will lead to Lotka–Volterra-like oscillations (see Weisberg & Reisman, 2008, for details). In fact, the oscillations are dependent on having something approximating *perfect mixing*, where the organism types are distributed evenly through the space. Because of the dependence of key model properties on spatial structure, when spatial structure is included in the model, then it is a focal property.

Here, then, is the problem: Theorists will imagine the spatial arrangement of predators and prey in different ways. Some of these ways will correspond to generators of the Lotka–Volterra oscillations, and some of them won't. But since the Lotka–Volterra oscillations are one of the key findings of investigations of this model, how can the imagined scenarios that don't correspond to them be instantiations of the model? It is unclear that the mutual belief principle could sufficiently restrict theorists' imaginations in this case.

The maths view offers a simple solution: The Lotka–Volterra model says nothing at all about spatial structure. It is a model about the coupling of population-level properties exclusively. This very fact of the model is, in part, responsible for some of the model's most distinctive properties. And whether this is a liability or an advantage depends on the purpose to which the model is being put. I don't believe that the mutual-belief principle generates anything like this simple solution.

Does the *de re* view also suffer from the problem of variation? It is not obvious that it does because the *de re* view suggests that instances of modeling are always tied directly to targets, with deviations from the target explicitly licensed by a model description. This is almost like a version of the reality principle, where the theorist starts with a real-world target and then is allowed to modify only what is specified in the description in the way that it is specified. But as I will argue in Section 4.4.3, the rigidity of this account that helps it solve the problem of variation generates problems of its own.

Although I have discussed in this section how the three accounts of models as fictions can respond to the problem of variation, I think the problem is still a significant obstacle for those who would defend such an account. Since focal properties of models look like they must extend beyond the text of the model description, if the view is to have any plausibility, proponents need to provide some principles of generation. On a Lewis-like account of the metaphysics, the model description attaches to an equivalence class of possible worlds which are fully instantiated in every way. This solution has the problem of generating too many properties if the entire equivalence class of worlds is taken to be the model, or else it simply magnifies the problem of variation, suggesting that theorists may not all be fixed on the same world and hence actually be entertaining different models.

Frigg avoids the problem of overgeneration of worlds' properties, but then has to explain how all of the needed focal properties can be generated in the first place. His own idea that laws of nature will generate the necessary properties is insufficient. A modified mutual-belief principle may be a partial solution, but will be very sensitive to individual differences between scientists. In the end, I don't know that the problem of variation is insurmountable for Frigg, but it certainly is a problem that must be addressed.

4.4.2 Representational Capacity of Different Models

A second problem for fictions accounts comes from apparent differences in the representational capacities of different models. As I discussed in Section 3.4, models can be discrete or probabilistic, aggregative or individualistic, spatially explicit or not, and so forth. If models are mathematical objects, these differences are easy to make sense of. Different kinds of models will use different kinds of mathematics and this will account for differences in their representational capacities. However, fictions accounts cannot make these distinctions.

Let's consider a specific example. Fictionalists regard the Lotka–Volterra model as an imaginary system composed of a predator population and a prey population. Setting aside, for a moment, how specific this has to be (are the predators sharks?), the model is composed of concrete, discrete organisms that interact with one another. But the equations used to describe the Lotka–Volterra model do so in terms of populations. This means that no individual organism is represented, in the mathematics, only populations of organisms. Does this matter?

To answer the question, let's consider some of the distinctive properties of the Lotka–Volterra model. It predicts that predator and prey populations oscillate indefinitely and out of phase, that the oscillations are neutrally stable, that its one equilibrium is unstable, and that a general biocide increases the relative abundance of the prey population.

Do any of these properties depend on the fact that the mathematics of the Lotka–Volterra model are population-level or that they are defined using differential equations, meaning that they are complete sets of trajectories through a state space? I think that the answer is clearly yes. When one takes an individual-based version of the Lotka–Volterra model, where individual organisms are represented discretely, but everything else is made as comparable as possible to the Lotka–Volterra model (which, admittedly, does involves some re-representation), then the properties change. For example, oscillations are no longer present unless density dependence is introduced. So this suggests that the mathematical representation changes the properties of the model, which fictionalists would have to deny.

In fact, even more subtle effects along these lines can be observed. Although it is analytically provable that the Lotka–Volterra model has undamped, fixed-amplitude oscillations, if one uses a computer to numerically approximate solutions to the differential equations, one will see that there is a very small increase in amplitude during each cycle of the oscillations. Numerical approximation routines introduce discreteness to the equations and, however small, these changes often affect their behavior.[9]

Much of the work on predation that followed Lotka's and Volterra's pioneering studies involved studying changes to the basic model of the sort described above. These changes in mathematics are reported as "changes to the model" in the literature. However, fictionalists cannot capture this practice in their understanding of the nature of a mathematical model. For a fictionalist, a model of predation has to be composed of concrete populations of discrete and distinct individuals. Whether the mathematics are population-level (as in the original Lotka–Volterra model) or individual-level (as in contemporary individual-based versions), the model is always composed of individuals. What theorists typically call differences in models (population vs. individual) will have to be called differences in the model description on the fictionalist view.

Although this is a possible solution, I think that claiming all differences reside at the level of the model description is very implausible. Since there are significant differences in the behavior of the model when the mathematics are changed, this should be accounted for by a difference in the model. Otherwise, we would have to say that some model descriptions (e.g., individual-based ones) are intrinsically better descriptions than others (e.g., aggregative ones). The specification relation between model description and model would be weak or absent in some cases.

Could proponents of the fictions view respond that there are imaginary populations that have only population-level, but not individual-level properties? I

---

9. It is tempting to say that numerical approximation *always* affects the behavior of such systems, but this is not quite correct. Special care can be taken to ensure that the method of discretization does not introduce unwanted affects by using symplectic or metasymplectic integrators. Such numerical approximation schemes can preserve certain invariants of the exact equations, insuring that key properties of the original equations are preserved after approximation. See Wisdom & Holman, 1991 for an example. I thank Glenn Ireley for pointing me to this literature.

am quite skeptical of this; this is certainly not something that I can imagine, although I admit that there may be variation in imaginative capacity among philosophers. Adopting a more metaphysically rich version of the fictions view might help here because on such a view imagination is not required, only the existence of these entities. Fictionalists could then posit imagined, concrete populations with no individual-level properties. However, given that such populations probably cannot be imagined, this sounds like special pleading.

A related problem about representational capacity has to do with probability. Many contemporary models in the physical, life, and social sciences are probabilistic at their core. For example, in standard implementations of the Schelling model of segregation, the initial distribution of agents to the virtual city is done by generating coordinates at random. In addition, many of the interactions have probabilistic elements. For example, in individual-based implementations of the Lotka-Volterra model, the initial distribution of organisms is generated at random by the computer.

It is unclear how such probabilistic interactions can be imagined or otherwise fictionalized. Any given fictional scenario will be a single instantiation of the probabilistic interactions. But how can a single instantiation actually represent the probabilities? One possibility is that we are supposed to imagine or otherwise entertain a fiction that is a composite of a large number of instantiations of all the probabilities. But if the probabilities are about things like "had food" or "didn't eat," or worse yet "lived" and "died," what is the composite supposed to be? Gambling predators and prey? It doesn't look like these probabilistic properties are expressible at the individual level because no individual can be both alive or dead, or satiated and hungry. An aggregate is expressible at the population level, but then we run into the problems discussed above and it is no longer clear that we are talking about the same model.

My overarching point here is that, while it is relatively easy to imagine the content of finite, deterministic, individualistic models like a population of genes undergoing assortment, it is unclear that this procedure could generalize to more complex cases. Aggregate models, infinite models, ensemble models, probabilistic models, high-dimensional models, and others can't be imagined in their entirety. This rules out the possibility of equating such models with imagined fictional scenarios and undercuts Waltonian versions of the fictionalist position.

4.4.3 Making Sense of Modeling

So far, we have considered the problem of variation and the problem of representational capacity as two major objections against fictionalist accounts of mathematical models. In addition to these two objections, Levy's *de re* account of mathematical models suffers from a problem that the other views can accommodate. Not only does Levy's view present a deflationary account of the nature of models, it also seems to present a deflationary account of the practice of modeling. Rather than positing mathematical modeling as the construction of fictional worlds, psychological states, or mathematical objects (*de novo* views), Levy argues that modeling is simply treating a target as if it had properties that it didn't have. While this certainly simplifies accounts of the model–world relation, it undercuts accounts about the practice of modeling as a distinct theoretical activity, which I think can be independently motivated. A consequence of Levy's view seems to be that there really is no difference between the practice of modeling and the practice of abstract direct representation (Weisberg, 2007b). Far from explaining the special uses to which models can be put, Levy says that there aren't any models at all.

Of course Levy might simply accept this consequence. He might argue that idealization and approximation are real things, but ultimately there is no such thing as modeling. Since I think that there are independent reasons for thinking of modeling as a distinct kind of practice, I would reject this line of response.

Another possibility is that Levy could argue, as I will in the next chapter, that the difference between modeling and other kinds of scientific representation has to do with the theorist's intentions, not with the kind of representation that is generated. When scientists model, they engage in a certain kind of activity. For Levy, this would mean that scientists adopt the presupposition that a target has some set of properties that it doesn't actually have. The problem with this response is that this kind of idealization is ubiquitous. Even in paradigm cases of direct representation, distortion is often introduced. So if he makes this response, Levy will be forced to claim that nearly all instances of theorizing are instances of modeling.

Finally, Levy's account gives us little guidance about more complex cases of modeling. Although I have been discussing cases where a single model is related to a single target system, this is not the only relationship between models and targets. In Chapter 7, I will introduce several other kinds of cases including the use of models to account for the properties of a generalized target and of a nonexistent target. In these cases, there is no real target, so it is unclear how the theorist can have *de re* beliefs about such a target.

4.4.4 Variation in Practice

My final objection to fictions accounts of mathematical models involves a reevaluation of the face-value practice. Recall that proponents of these accounts point to many convincing examples where scientists seem to rely on imagining fictional scenarios in order to introduce and think about their models. I believe that this is part of some scientific practice and will suggest a way that we can account for it in Section 4.5. However, despite accepting the presence of this practice, I think its prevalence and its importance have been overstated. For all the examples that can be given where theorists seem to invoke a fictional scenario, there are also many where they do not invoke a fictional scenario, or at least don't do so in any straightforward way.

Let's begin with a case from population biology that is similar to the example from Maynard Smith discussed above, but in which an imaginary population isn't referred to. In Karlin's and Feldman's discusion of the unsymmetric equilibria in cases of loose-linkage disequilibrium, they begin their introduction of the model with the following text:

Let x_1, x_2, x_3, and x_4 be the frequencies of the chromosomes AB, Ab, aB, and ab, respectively, and r the recombination fraction. Then corresponding to [equation 1], the recursion relations relating the frequencies x_1', x_2', x_3', and x_4' in the next generation to x_1, x_2, x_3, and x_4 have been shown to be:

$$\bar{w}x_1' = x_1 - \delta x_1^2 - \beta x_1 x_2 - \gamma x_1 x_3 - rD$$
$$\bar{w}x_2' = x_2 - \beta x_1 x_2 - \alpha x_2^2 - \gamma x_2 x_4 - rD$$
$$\bar{w}x_3' = x_3 - \gamma x_1 x_3 - \alpha x_3^2 - \beta x_3 x_4 - rD$$
$$\bar{w}x_4' = x_4 - \gamma x_2 x_4 - \beta x_3 x_4 - \delta x_4^2 - rD$$

where $\bar{w} = 1 - \delta(x_1^2 + x_4^2) - \alpha(x_2^2 + x_3^2) - 2\beta(x_3 x_4 + x_1 x_2) - 2\gamma(x_1 x_3 + x_2 x_4)$, and $D = x_1 x_4 = x_2 x_3$ is usually called the disequilibrium value. (Karlin & Feldman, 1969, 70-71)

In this example, Karlin and Feldman do not appeal to any particular, concrete population. Rather, the entire mathematical argument is made in terms of extremely general properties of infinite populations. These properties might be said to abstract over all real and imagined populations, but no reference is made to these populations. In fact, the arguments may not even make sense if any particular population was being referenced. Nor do Karlin and Feldman give any indication that they are imagining a specific population as a cognitive aid.

There are also cases of mathematical modeling that are even more remote from anything concrete and imagined. Say that we were investigating an approximate quantum mechanical model of a chemical reaction. Such a model is constructed by taking into account the forces acting on the molecular system and giving an approximate accounting of all of these forces. The resulting model takes the form of a set of paths through a potential energy surface. The space itself is of high dimension ($3N - 5$ dimensions, where N is the number of atoms in the molecule). The paths through this space cannot be thought about or imagined. They have little resemblance to the concrete properties of physical molecules, except for the correlation of potential energy to molecular coordinates.

Models can be even more abstract than the cases discussed above. The standard models in sciences like statistical thermodynamics quantify over ensembles of states, which are probability distributions over probability distributions of states. This is even more remote from anything that can be imagined, compounding the problem of imagining complex spaces with probabilities. These cases illustrate that, in addition to any variation in scientists' capacity for imaginative cognition, some model systems are simply unimaginable.

Defenders of fictions views have several possible lines of response to these variations in practice. First, they could simply argue that, despite appearances, all mathematical modeling involves imagined systems. A molecular model really is about imaginary molecules whether or not the theorist is capable of mentally connecting the mathematics to anything concrete. I don't see how this argument could be made if one adopts the Walton-style metaphysics. On this type of view, scientists would actually have to be aware of the fictions they are entertaining because they are constructing the fictions by entertaining them. Lewis-style metaphysics allows for the possibility of unimagined possibilities, although accepting unimagined possible worlds as the model seems to undermine the face-value practice as a justification for the fictions view.

A more plausible response is for fictionalists to simply adopt a pluralist stance. They could acknowledge that practices vary and that in some cases, interpreted mathematics alone is sufficient for theorizing. This would weaken their position, but still allow them to speak about many core cases as fictions. If they take this position, then we might ask how differences in cognitive style will matter to the account. Perhaps some scientists imagine their models most of the time, some little of the time. Some may never imagine scenarios for their models, and there may be models associated with unimaginable scenarios.

While I certainly think a kind of pluralism is called for here, I think that the kind of fictionalist response discussed above doesn't capture it correctly. What we should be pluralistic about are the cognitive abilities of scientists. This pluralism suggests that there are many different ways to get a handle on mathematical models and not all of them involve the imagination. Thus we should look for a "supporting role" for the imagination and should not put it at the center of an account of mathematical models and modeling.

Let's take a step back and review my four arguments against the fictions view. As in my description of the practice of modeling itself, I think it is useful to work at the epistemic level, where this means that what we are after is the best reconstruction of scientists' cognitive practices (see Section 2.6). Followed strictly, this would suggest that matters of ultimate ontology are irrelevant to the discussion here. But things can't be kept that clean because at least some ontological issues are relevant to responding to challenges against this view.

To take one example, the problem of interscientist variation can seem like more or less of a problem depending on questions of a metaphysical character.

On a Waltonian account, where fictions rest on imagination, interscientist disagreement about focal model properties is a major concern and it is unclear what kind of machinery can solve this problem. This is because Waltonian accounts identify the model with the fiction itself. Conversely, Lewis-like metaphysics can solve this problem because no particular scientist's imagination is required to individuate models. However, accounts based on the metaphysics of possible worlds have a hard time accounting for scientists' epistemic access to their models. How can they account for possible worlds which scientists have created but cannot imagine?

Maybe it is possible to separate the epistemic aspects of the Walton or Lewis machinery from any substantial metaphysical commitments. That would be ideal methodologically, but extant literature has not revealed how this can be done. So we are faced with a situation where at least some issues of metaphysics cannot be screened off from accounts of models and modeling.

The considerations raised in the proceeding sections further suggest that the invocation of fiction is not a simple way of dealing with puzzling aspects of the practice of mathematical modeling. And if my discussion in this section is accepted, it is hard even to understand how the case can be made that fiction is required for mathematical modeling because there are models that seem impossible to translate into fictional scenarios. However, I am not trying to argue that imagined scenarios are never helpful or even that they have no cognitive role in science. In fact, I will now turn to the precise ways in which I think imagination can aid the practice of modeling.

4.5 ■ FOLK ONTOLOGY

Although I have argued against the fictions account of mathematical models, I believe that the face-value practice of modeling does include frequent appeals to fictional scenarios. Imagining concrete phenomena seems to guide some theorists as they construct and analyze mathematical models. This practice ought to be accounted for, even if one doesn't accept the fictions account.

Recognizing that the metaphysics of concrete but imaginary systems is complex and by no means settled, Godfrey-Smith suggests that we needn't be too committal about the metaphysical status of mathematical models. He writes:

> ...to use a phrase suggested by Deena Skolnick [Weisberg], the treatment of model systems as comprising imagined concrete things is the "folk ontology" of at least many scientific modelers. It is the ontology embodied in many scientists' unreflective habits of talking about the objects of their study—talk about what a certain kind of population will do, about whether a certain kind of market will clear. (Godfrey-Smith, 2006, 735)

This is a good suggestion and I think we should take it literally. I propose to treat theorists' imagination about fictional scenarios as their folk ontologies, which exist alongside the model and its construal. So what exactly is the folk ontology of a model and how does it operate?

Let us look at how the Lotka–Volterra model is presented in Maynard Smith's important monograph on ecological theory. Several sections after he introduces Volterra's model of predator–prey relations, he asks us to suppose that "some number $[V_r]$ of the prey can find some cover or refuge which makes them inaccessible to the predator" (Maynard Smith, 1974, 25). This gives us a model which is described by the following differential equations:

$$\frac{dV}{dt} = rV - aP(V - V_r) \qquad (4.1)$$

$$\frac{dP}{dt} = baP(V - V_r) - mP \qquad (4.2)$$

He goes on to explain how this new model makes two interesting predictions. First, if the total number of prey in cover is a constant fraction of the total, this does not alter the nature of the oscillation and unstable equilibrium. However, "if the number of prey in cover is constant ... the effects of cover are stabilizing, since it changes a conservative into a convergent oscillation." In other words, a constant number of prey hiding from predators stabilizes the oscillation of the model.

The details of this model and its comparison to the original Lotka–Volterra model are interesting, but what I want to focus on is the use Maynard Smith made of concrete imagery. He began by telling us to imagine a Lotka–Volterra predator–prey system. He then gave us some information about how to modify our imagination: We are to imagine that some fraction of the prey population was allowed to find cover, avoiding the predators. No doubt any reader would find herself imagining some population of prey heading for cover. Of course, huge variation among scientists exists. I first imagined sharks and salmon, since I was thinking about Volterra's models. Others may have though about a lizard hiding under a rock from a kookaburra. Still others may have thought of clownfish, which hide in anemones. On the fictions view, these mental pictures (or possibly some abstraction of them) are actually the model. I think that these mental pictures are *aids to thinking about the model*, but are not part of the model itself.

I follow Deena Skolnick Weisberg in calling these mental pictures the folk ontology of models. Just as folk psychology helps people make predictions about the behavior of others, and folk thermodynamics helps people figure out what will be too hot to touch, folk ontology aids theorists in developing mathematical models and the equations that describe them. And just as folk psychology will undoubtedly vary from person to person, but be in near

Fictions and Folk Ontology ■ 69

enough agreement to make predictions, so folk ontology can vary among theorists.

Godfrey-Smith and Weisberg are correct that many theorists describe imagined concrete systems when they are talking about models, but I advocate interpreting this talk as a commitment to folk ontology, not to models being just what theorists imagined. Given this view, it is natural to ask whether folk ontology is an essential part of the practice of modeling or is it something that can be completely dispensed with. I believe that the former is correct; folk ontology is essential to modeling in at least three contexts.

The first role of folk ontology is to assist in the development of a mathematical model. Take the initial formulation of Volterra's predator–prey model as an example. We don't have access to Volterra's mental representations, but it is probably fair to say that he began by imagining a population of predators and a population of prey and attributed to them certain properties. Because he wanted to perform a mathematical analysis of this population, he set this idea to paper, writing down equations which specified the model that he had imagined. We do not have a record of this, so we do not know how satisfied Volterra was with the initial model. Perhaps it did not match the system he had imagined and so he refined the model. Or perhaps he had correctly specified the model he was imagining and was able to proceed to analyze it. In either case, the mental picture he had—his own folk ontology of the model—guided him in formulating the initial model description and making sure that these equations picked out what he had in mind. So one role of a theorist's folk ontology is to guide the development and refinement of a mathematical model.

The second context in which folk ontology is important is in thinking about very complex mathematical models. Consider some of the complex mathematical models employed in chemistry. Even highly idealized models of the reactions of simple molecules consist of potential energy surfaces in state spaces of high dimensionality. No chemist can hold this picture in her mind and hence cannot directly reason about the imagined system. All she can do is manipulate it on the computer. However, she does have access to a mental picture which, more or less, corresponds to the assumptions and idealizations of the model. This need for such mental pictures is all the more dramatic in statistical thermodynamics. Statistical mechanical models of gases are actually ensemble models. They are "a hypothetical collection of an infinite number of noninteracting systems, each of which is in the same macrostate (thermodynamic state) as the system of interest" (Levine, 2002, 749). Such an ensemble is not something that can be thought about directly; highly approximate mental pictures are the only ways to think about these systems.

Finally, the folk ontology of mathematical models plays an important role in coordinating models encoded in different representational systems. These models might be very different mathematically, but they nevertheless share many features or assumptions in common. For example, predator–prey models are most commonly developed in an aggregate way, where the main quantities tracked are populations of organisms. However, the contemporary ecological literature has increasingly turned to individual-based approaches, where each organism is represented explicitly as an individual (Grimm & Railsback, 2005).

Very similar models can be developed in the individual-based and aggregate frameworks in the sense that these models have the same kinds of feedback loops. Yet such models are mathematically distinct because, for example, the state spaces of individual-based models will often have hundreds more dimensions than the state spaces of aggregate models, since their dimensionality scales with the number of organisms. On the mathematical account of mathematical models, then, these will be distinct kinds of models. On the concrete account, they can potentially be the same model, described using different mathematical language.

Neither of these pure perspectives is satisfactory. Aggregate and individual-based models of the same phenomenon clearly have significant differences in their structural properties. Yet to say that they have no more than a superficial relationship with each other seems too strong. Here is a place that the folk ontology of our models can help. Folk ontology lets us tie the very complex behavior of the individual-based model back to the simple aggregate model. It can function in a similar way whenever theorists need to compare topically similar but structurally different models.

For all of these reasons, theorists' folk ontologies about their models seem to be a crucial part of scientific practice. Without these mental pictures, it would be difficult to develop mathematical models in the first place, to hold mathematically complex models in mind, and to coordinate similar models embedded in different representational systems. A full account of models must include the role of these mental pictures, but they are not, I believe, most appropriately thought of as the models themselves.

4.6 ■ MATHS, INTERPRETATION, AND FOLK ONTOLOGY

We come now to a complete picture of mathematical models, concluding the discussion of the last two chapters. Mathematical models are mathematical objects, described by model descriptions. Many different kinds of mathematical objects can serve as mathematical models, although they are often trajectory spaces. The model's structure is interpreted via the theorist's construal, which determines what each dimension of the state space means. When a model is intended to be compared with a real-world phenomenon, the nature of this phenomenon is specified in the intended scope, another part of the theorist's construal. Comparison between model and target, the subject of Chapter 8, is assisted by the theorist's folk ontology and is evaluated according to standards set by the theorist's fidelity criteria.

With a complete account of scientific models sketched, I want to highlight the features of this account that play the most important roles in the practice of modeling. I will outline three major themes.

First, the account of mathematical models I have laid out draws a strong parallel between mathematical models and concrete, physical models. Both concrete and mathematical models stand in many-to-many relationships with their model descriptions, require interpretations to be fully realized, and can potentially stand in many different kinds of model–world relations with target systems. Most of the modern literature about scientific models pays little attention to concrete models, as their use has become considerably less important with the rise of computational methods. Nevertheless, I believe that the strong parallels between these accounts will motivate further research into the representational capacities and fruitful uses of concrete models. In this respect, I share much in common with proponents of fictions views, who often emphasize the parallels between mathematical and concrete models.

More central to the concerns of extant accounts of modeling practice are the many roles for theorists' intentions that I have discussed in this and the last chapter. Theorists' intentions play a role in determining what counts as a model, how the model is individuated, which aspects of the world are to be parts of the target, which bits of the model represent which bits of the world, and what standards of fidelity are used to evaluate the model. If we are to understand the ways in which modeling can be conducted most effectively, then much more needs to be known about the kinds of decisions that can be made about these factors and what effect these decisions have on the representational power of models.

Finally, I have argued that the folk ontology of mathematical models, the beliefs and mental images that individual and communities of scientists associate with the abstract mathematical object, play several indispensable roles in modeling. If this is correct, then traditional attitudes about the unimportance of psychology and pragmatics to philosophy of science (e.g., those in Hempel, 1965) can have no place in detailed accounts of models and modeling.

I opened this chapter with a discusion of the major reasons philosophers give for rejecting maths accounts and developing fictionalist accounts of models. These reasons involve individuating models, the need for mathematical models to have causal structure, and the face-value practice of mathematical modeling. We can now consider how my account would respond to each of these charges.

The individuation objection is that the very same mathematical structure is used in what look like distinct models. The harmonic oscillator equation is used in models of springs, pendula, and molecular vibration. But if a mathematical model is just a bit of mathematical structure, how can we treat these as different models?

This objection is the easiest one for proponents of maths accounts to address. As I argued in Chapter 3, mathematical models are not exclusively mathematical structures. Rather, they are *interpreted* mathematical structures. So when the same mathematical structure is used to describe a pendulum, a spring, and a circuit, what differs between these models is the interpretation of the structure, especially the assignment and the theorists' intended scope. In fact, fictions accounts have their own problem of individuation: they cannot account for what such models have in common when they share the same mathematical structure. The best that they can do is to say that these models are described by the same mathematics, but not that they have any core structure in common. However, on the maths account, we can say that the models share a common core mathematical structure, and what differs are theorists' construals.

A second objection to the maths accounts is that mathematical structures are not causal, but scientific models often convey causal information. How can models convey causal information if they are mathematical objects?

Not all models are intended to be causal, but when they are, causal structure is represented by the transition rules associated with the model. These will take the form of functions in a mathematical model or algorithms in a computational model. These structures alone don't make the model causal, so we need to look beyond the model's mathematical structure to its construal.

For example, the Lotka–Volterra model is intended by ecologists to be interpreted causally. The coupling between the two differential equations is intended to represent a causal coupling between two populations of organisms. Mathematical couplings are not causal, and the fact that they can represent causal couplings is not solely a matter of their mathematical structure. There is no mathematical reason that a differential equation should represent causal structure. Rather, it is the use to which these equations are being put that determines whether they are representing causal relations or not. Hence, it is imperative to consider the construal, specifically the assignment, when determining whether one is dealing with a causal model.

Once a theorist knows that she is working with a causal model, further determinations of representational adequacy can take place. If she has determined, for example, that the coupling structure of the differential equations is supposed to represent causal coupling in the world, then she can endeavor to determine empirically whether or not the target is so composed. Similarly, if the equations describing the model have an additive structure, where various variables are linearly combined to determine the value of another, she can determine whether this reflects a real causal relationship in the target, possibly through interventionist experiments.

How is the causal structure of the real system supposed to be determined such that it can be compared to the model? This is an extremely difficult question and I will say more about it in Chapter 5. But a growing consensus in the literature on causation is that counterfactual relations are a core element of causal relations (Woodward, 2003; Collins, Hall, & Paul, 2004; Schulz, Gopnik, & Glymour, 2007; Strevens, 2008). Scientists explicitly intervene on and control variables so that they can study the effect that variables have on one another. Something similar can be done with models themselves: Parameters can be changed, ranges of initial values can be tested, and new couplings or functions can be introduced to models. In each case, the theorist studies how these changes affect the model's behavior.

So one way that modelers investigate the causal properties of a real-world system and try to ensure these are represented by the model is to plan parallel experiments between target system and model. As much as is feasible, they manipulate properties of the target system and see the systematic effects of these manipulations. They can then conduct parallel manipulations on the parts of the model intended to represent those parts of the target system. Converging results between model and target system are evidence that the model is capturing the real causal structure of the system.

I should note, once again, that a modeler's fidelity criteria set the standards for how good a match there must be between the model's causal structure and the world. Sometimes, modelers may not care at all about capturing causal structure; perhaps only the output matters. In other cases, they may be content to have the primary causal factors of the target captured in the model. And in other cases, theorists may be after a very precise and accurate model that can not only make extremely high-fidelity predictions, but that also captures, with high fidelity, the causal structure of the target system. In these cases, an interpreted mathematical structure (i.e., maths plus construal) is perfectly capable of representing a causal structure in the target system, and theorists have techniques for assessing the fidelity of the representation.

The final problem is how we can account for the face-value practice of modeling. My answer has two parts. First, I think that there is no single face-value practice of modeling. Some theorists some of the time (or even many theorists much of the time) appeal to fiction when thinking about their models. This is an important observation about scientific discourse and cognition. However, all of the cognitive roles that fiction plays in the practice of modeling can be accounted for by folk ontology, and this doesn't require claiming that all mathematical models are fictions. Treating appeals to fiction as part of folk ontology has the added benefit of showing why these appeals are optional in some cases: Fictions sometimes play a role in representation, but are not the only things than can do so. If their core role in modeling were to serve as potential representers, then this variation in practice would be impossible. My complete picture of models as being composed of structure and construal, and being accompanied by folk ontology, is thus the best way to account for the face-value practice.

5 Target-Directed Modeling

One of the virtues of modeling, as opposed to other theoretical practices such as abstract direct representation, is that it is extremely flexible. Models can be used to study a single target, a cluster of targets, a generalized target, or even targets known not to exist. One can even engage in the study of a model without any target at all. In this chapter, I will consider the simplest case of modeling, what I call *target-directed modeling*. This type of modeling requires the modeler to have a specific target system in mind and to generate predictions and explanations about this specific target. Not only is target-directed modeling the simplest type of modeling, it is also the basis of the more complex cases I will discuss in subsequent chapters.

One example of target-directed modeling that I have already discussed several times is Volterra's study of the Adriatic fisheries. Describing how he conducted his research, Volterra wrote:

> As in any other analogous problem, it is convenient, in order to apply calculus, to start by taking in to account hypotheses which, although deviating from reality, give an approximate image of it. Although, at least in the beginning, the representation is very rough ... it is possible to verify, quantitatively or, possibly, qualitatively, whether the results found match the actual statistics, and it is therefore possible to check the correctness of the initial hypothesis, at the same time paving the way for further results. Hence, it is useful, to ease the application of calculus, to schematize the phenomenon by isolating those actions that we intend to examine, supposing that they take place alone, and by neglecting other actions. (Volterra, 1926b, 5, translation G. Sillari)

Volterra describes a procedure by which one constructs a model by first making some assumptions about the causal structure of the system of interest. One then "applies calculus," writing down equations describing the constructed model. Only after one understands the consequences of the model, what he calls the "actions we intend to examine," does one try to compare the model to one's real-world target. This corresponds very well to the three distinct elements of target-directed modeling: development of the model, analysis of the model, and targeting the model to a real-world system. These are conceptually distinct, but in practice, may happen together or even iteratively. Complete instances of target-directed modeling have all three of these elements, even though some of them may be borrowed from other research episodes and may only be implicit. This chapter will consider the three elements in turn.

5.1 ■ MODEL DEVELOPMENT

The first element of target-directed modeling is the development of the model itself. I use the term 'development' for this aspect, because it is very much an active process. Theorists must either construct or borrow a structure, evaluate the structure's representational capacity, and develop a construal for this structure.

Constructing a model's structure can take a number of different forms. In the case of concrete models, literal construction is involved and this can take considerable time and engineering skill. In mathematical modeling, construction is achieved by writing down the model description for the model, typically in the form of equations or graphs. Computational modeling is similar; a procedure is specified using natural language, discrete mathematics, pseudocode, or a programming language.

In some instances of modeling, construction is carried out from scratch. For example, when Volterra first constructed the mathematical structure for the Lotka–Volterra model, he had no previous biological models from which to work. However, in many cases, structures are developed from other structures, or may be completely coopted from another use. Any subsequent researchers using the Lotka–Volterra model are borrowing their mathematical structure from Volterra and from Lotka.[1]

Theorists don't construct or borrow any old structure. They aim to choose a structure that has an adequate representational capacity for their chosen target. For example, Schelling could not have used standard, decision-theoretic mathematics for his model of segregation, because it is difficult, if not impossible, to represent spatial positions and relations with that mathematics. Instead, he chose a structure that was adequate for representing space and the proximity of agents to one another (see Section 3.4 for further discussion of representational capacity).

Determination of representational capacity can be a formal affair, and is sometimes necessary to verify the predictive adequacy of a model. In such cases, this determination usually happens after a model is fully constructed. But a less formal determination of representational capacity is almost always required in

---

1. A detailed analysis of the process of elaboration of models, from simple and highly idealized to more complex and realistic, can be found in Wimsatt (1987).

the early stages of constructing a model, in order to make sure that further work will be fruitful.

Once the model's structure has been constructed or borrowed, and its representational capacity has been established, the theorist must construct a construal for the structure. Recall that construals are composed of four parts: an assignment, the modeler's intended scope, and two kinds of fidelity criteria (Section 3.3). The assignment and scope determine and help modelers to evaluate the relationship between parts of the model and parts of real-world phenomena. The fidelity criteria are the standards that theorists use to evaluate a model's ability to represent real phenomena.

I have already explained the function of the parts of the construal in Chapter 3, so here I want to focus on what it takes to establish these parts. Specifically, exactly what does it mean for a theorist to have a construal? Is it something that must be written down? Does it need to be made explicit? Can it change in the course of modeling?

As with all manner of interpretations, theorists' construals can vary widely in their explicitness. At one extreme is a formal treatment of a model, where every part of the model's structure is either named or explicitly ignored, the intended scope of the model is clearly delineated, and the standards of fidelity are stated precisely. One can find these kinds of completely explicit construals in at least two kinds of cases. The first is in formal, after-the-fact treatments of successful models of the sort one finds in rigorous textbooks (e.g., Kittel, 1980). The second is when a particular model or family of models is very controversial for a given target. Conflict about the nature and scope of a model gives incentives for theorists to become extremely explicit about the nature of their models and the standards against which their models are being evaluated.

These kinds of cases, however, are exceptions to the rule. More often than not, one or more elements of the construal are not made verbally or formally explicit, leaving these elements open to differing interpretations. But if every theorist could potentially have their own construal for their model, how can these be shared by a community of scientists?

While I think that this situation of massively differing construals is possible, it doesn't usually happen because of a scientific community's shared background knowledge. This background knowledge imposes significant constraints on construals. If this is correct, one should expect this knowledge to be considerably more explicit early in theoretical inquiry, becoming more implicit as the practice develops. While I haven't done a systematic study to this effect, when one peruses published theoretical articles, it is very clear that construals are often left implicit in this way. As an outsider, it can be extremely difficult to understand the significance of many papers, including their intended scopes and standards of fidelity.

Does this mean that the construal lies completely with the interpretive intentions of the community, instead of individual scientists? Again, I think there is considerable variance. In some cases, construals are so widely accepted in a community that they never need to be explicitly communicated. A theorist can simply write down a model description, or build a concrete model, and the construal is automatically assumed by the theorist and the model's potential audience. In other cases, there are widely accepted standard construals, but the theorist wants to introduce deviations from these standards. This requires more explicitness, of course. In still other cases, perhaps the most common ones, construals will be partially constituted by community-level beliefs and partially constituted by the implicit or explicit intentions of the theorist.

Thus far I have talked about various aspects of the construction of a single construal for a model structure. The last issue I want to discuss in this section is the extent to which construals can change in the course of modeling. It turns out that there can be quite substantial change of construals, even during a single episode of target-directed modeling. More dramatically, the same structure can be given an entirely new construal when a model is borrowed from another area or problem.

Why would a construal change during a single episode of target-directed modeling? One reason, especially in early, exploratory stages, is that theorists simply don't yet "have a feel" for their models. Since modeling involves constructing and analyzing models, not the extracting patterns from data sets, it may be very difficult to know, for example, how much fidelity could reasonably be expected of a model. So one might start with high hopes and extremely high levels of fidelity, but a bit of manipulation of the model might suggest that the best that can be achieved is a qualitative fit to the target.

Another, related reason is that the model structure has properties not initially appreciated, and some of these properties are not considered desirable for the model. To take a simple example, if it is determined mathematically that a model has unrealistic equilibria in the limit, but is "well behaved" on shorter timescales, theorists might choose some kind of temporal restriction on the model. The opposite issue can also arise. A structure might have particularly desirable properties initially ignored, but these features are later included in the model's assignment as their significance is appreciated.

The most dramatic examples of construal changes for a structure come when the entirety of the assignment changes. A very nice example of this can be found in Richard Goodwin's reinterpretation of the mathematical structure of the Lotka–Volterra model for use in economics. Rather than give a set of coupled differential equations that describe the fate of two populations, Goodwin sets up the model in terms of two macroscopic economic variables: u, the workers' share of the national income, and v, the employment rate.

After stating a number of assumptions and assigning the key variables of the model description, Goodwin writes:

From this ... we have [the following] convenient statement of our model.

$$\dot{v} = [1/\sigma - (\alpha + \beta)) - 1/\sigma u]v$$
$$\dot{u} = [-(\alpha + \gamma) + \rho v]u$$

In this form we recognize the Volterra case of prey and predator [...]. To some extent the similarity is purely formal, but not entirely so. It has long seemed to me that Volterra's problem of the symbiosis of two populations—partly complementary, partly hostile—is helpful in understanding the dynamical contradictions of capitalism, especially when stated in a more or less Marxian form. (Goodwin, 1967, 55)

By reinterpreting the Lotka–Volterra structure in this way, Goodwin was able to construct a model of employment and workers' wage share. This model predicts that high unemployment can generate wage inflation. For a given set of parameters, the basic result is that increases in the employment rate are negatively coupled to increases in the workers' share of the national output. As employment increases, the workers' share decreases. But when the wage share decreases below a certain threshold, this starts driving the employment rate down, which in turns starts driving the workers' share up again. Like the Lotka–Volterra model, there is one unstable equilibrium ($\hat{v}$, $\hat{u}$) for this model, which will correspond to the time average of v and u.

In his own description of the model, Goodwin noted its relationship to the Lotka–Volterra model. He says that this relationship is "not entirely formal," but what does this mean? The best way to parse this phrase is as making three distinct claims. First, Goodwin recognized that the two models are entirely similar formally because they share the very same mathematical structure. Second, the models are not identical because the construals are different. Volterra assigned P and V (u and v) to predator population size and prey population size, while Goodwin assigned them to the workers' share of output and employment rate respectively. Finally, once fully interpreted, the models share a higher-order similarity. There is an analogy between the predatory relationship where predators drive down the population of prey and the way that high wages can drive down the employment rate. As an economist sympathetic to Marx, Goodwin was prepared to take the analogy of predation quite seriously.

Goodwin's use of the Lotka–Volterra structure is an especially clear illustration of how the same structure can become a different model with a different construal. It is also a bit of an unusual, extreme case, in which a mathematical structure is completely reinterpreted in a new domain. There are many less extreme cases of construal change, which are probably more common, but no less interesting. For example, some of the core models of game theory were first formulated in economics, further developed in biology, then reimported back into economics as a body of models called evolutionary game theory (Grüne-Yanoff, 2011). In the other direction, models of natural selection have been imported into economics (Rice & Smart, 2011). And the examples could be multiplied. Many physical models have found a home in ecology. Chemical models borrow structure from physics, and give structure back to political science and sociology, and so forth.

What we have seen in this section is that the first aspect of modeling, constructing the model, is a process with two parts. The first part requires that a structure has to be constructed or acquired from elsewhere, and the second part requires that a construal must be fixed for the model. The construal may be very explicitly stated, or remain implicit. This construal can change through time, or with the application of the same structure to a different modeling domain.

Once the structure and construal are at least temporarily fixed, the creation of the model is complete. Its structure is now meaningful and it can stand in relations of denotation to real and possible target systems. The second element of target-directed modeling involves putting the model to work.

5.2 ■ ANALYSIS OF THE MODEL

The second element of target-directed modeling is the analysis of the model. This is the core of target-directed modeling, where much of the effort of modelers is applied. Model analysis takes many different forms depending on the type of model, the interests of the scientist, and pragmatic factors including time, available computational power, and so forth. Nevertheless, there are some general characteristics of target-directed modeling's analysis component and I will begin by discussing these. I will then detail some of these points by considering the Bay model, the Lotka–Volterra model, and the Schelling model.

5.2.1 Complete Analysis

One way to start thinking about model analysis is to think about the goal of this part of the practice. In some cases, a modeler will have the goal of *complete analysis* of the model. Achieving this goal means that a theorist will know, or have a representation of:

1. the static and dynamic properties of the model;
2. the allowable states of the model;
3. the transitions between states allowed by the model;
4. what initiates transitions between states;
5. the dependence of states and transitions on one another.

I will refer to this list as the *total state* of the model.

For example, in a dynamical model like the Lotka–Volterra model, a complete analysis will give all of the possible coexisting population abundances for the two species, the transitions among these states, the stable and unstable equilibria, the amplitudes of neutrally stable oscillations, and so forth. Typically this analysis would be performed in such a way that models generated with differing parameter sets could also be examined.

A complete analysis of a concrete model like the San Francisco Bay model would look very similar. It would involve the determination of the model's behavior under a particular set of settings for its appurtenances. Ascertaining the directly measurable states, such as local velocity, tidal elevation, and salinity through time, are part of the complete analysis. So is ascertaining the more abstract properties such as the values of dimensionless quantities like the Reynolds, Weber, and Cauchy numbers for the system.

The goal of complete analysis is to know the total state of the system, but how is this determined? In the simplest cases, such as when we are dealing with models described by linear, first-order differential equations, we learn the total state of the system analytically using mathematics. To take a very simple example, consider the exponential growth model in ecology. This model is described by the following differential equation

$$\frac{dN}{dt} = rN \qquad (5.1)$$

where N is the population size and r is the growth rate of the population.

This equation can be solved for $N(t)$, the size of the population at a given time, by separation of variables yielding:

$$N(t) = N_0 e^{rt} \qquad (5.2)$$

where N_0 is the initial population size.

Equation 5.2 gives us a compact way of stating the complete analysis of the model. It says that a population of size N_i will grow exponentially as a function of time. The shape of the exponential curve will depend on the growth rate r, with positive growth rates leading to exponential increase, negative growth rates leading to extinction, and zero growth rates leading to no change. The model says nothing else at all about the change in population sizes over time. Having this complete story about the model's dynamics in a compact form is what it means to have an *analytical solution* to a model.

More complicated models, which are described by more complicated model descriptions, do not lend themselves to such simple analyses. For example, since the Lotka–Volterra model is described by two coupled differential equations, we cannot generate an analytical solution describing its complete behavior.

However, much can be learned about the model by mathematical analysis of its structure.

We can begin by investigating whether or not the Lotka–Volterra model has any nontrivial equilibria. This can be done by setting both equations to zero and solving for V and P. We find that there is one equilibrium with values:

$$\hat{V} = \frac{m}{ab} \tag{5.3}$$

$$\hat{P} = \frac{r}{a} \tag{5.4}$$

This equilibrium is unstable, but corresponds to the average abundance of each of the species.

Another question concerns the stability of the oscillations: will they continue indefinitely, or will they dampen down to some equilibrium population sizes? This can be analyzed by constructing the community matrix for the model. To do this, we start from the (unstable) equilibrium point and write down the community matrix, which has the following form:

$$A = \begin{pmatrix} 0 & -\alpha m/\beta \\ \beta r/\alpha & 0 \end{pmatrix} \tag{5.5}$$

If we solve for the eigenvalues of this matrix, we get the complex conjugate:

$$\lambda = \pm i(\alpha\beta)^{1/2} \tag{5.6}$$

Because this eigenvalue has real parts equal to zero, the oscillations will be neutrally stable. This means that they will continue indefinitely and, if they are disturbed in any way, they will not have a tendency to return to their original amplitude, nor will they become unstable (May, 2001, 42).

This analysis of the Lotka–Volterra model cannot be expressed as a single equation, but it is still complete. We can learn about the behavior of the model over its entire domain by doing mathematics. In more complex mathematical models, algebraic analysis may no longer be available. In such cases, theorists resort to a set of tools collectively called *numerical analysis*. All of these tools use some kind of numerical approximation method, instead of algebra, in order to analyze models. Humphreys explains this type of analysis as follows:

> In asserting that an equation has no analytic solution we mean that the function which serves as the differand has no known representation in terms of an exact closed form or an infinite series. An important feature of an analytic solution is that these solutions can be written in such a way that when substituted into the original equation, they provide the solution for any values of the variables. In contrast, a numerical solution gives the solution only at finitely many selected values of the variables. The switch from analytic mathematics to numerical mathematics ...

has one immediate consequence: We often lose both the great generality and the potentially exact solutions that have traditionally been desired of scientific theories. (Humphreys, 2007, 65)

Sometimes, such analysis can generate a general picture of the behavior of the model over its entire domain. Other times, numerical analysis involves *simulation*, computing the behavior of the model using a particular set of initial conditions (Winsberg, 2010), which gives the result only for that set of initial conditions. In such cases, a complete analysis of the model requires many simulations to study the behavior of the model over its entire domain.

Say that you had adopted the ideal of complete analysis, and hoped to understand general properties of a model's behavior. Looking at one set of initial values for the model would not be nearly enough to generate a complete analysis, and yet it would be difficult to produce even this much. In such cases, theorists experiment with their models, sampling the behavior in particular regimes in order to make inferences about the behavior as a whole (Humphreys, 2007; Winsberg, 2010). For example, imagine a model whose description had ten independent variables, and for which there was no possibility of an analytic or numerical solution. In such a case, the only way to give a complete description of the behavior of the model would be to try all possible values for the ten variables and compute the behavior of the model. This is often prohibitively time-consuming or even impossible. Instead, theorists can draw values of the variables from a known probability distribution and then compute the behavior of the model for those variables. Statistical techniques can then be used to infer the behavior of the model in the domains not explicitly computed. These types of technique are often called Monte Carlo methods, after the famous casino in Monaco.

One of the interesting things about Monte Carlo methods is that they are almost exactly analogous to how one would have to go about studying the complete properties of a physical model. If you really wanted to understand the entirety of the static and dynamic properties of something like the Bay model, you would have no choice but to treat it like any other physical system: take measurements and then infer the values that were not directly measured. This parallel and others has led some to claim that many simulations are themselves kinds of experiments (Guala, 2002; Morrison, 2009; W. S. Parker, 2009; Winsberg, 2010; but for the opposite view see Giere, 2009).

As models become even more complex, direct computation for even a single set of initial conditions may not be possible. For example, the physical processes that give rise to the weather are well known, and extremely accurate models for these processes can be derived. However, when these processes are combined together and then applied to such a large system as Earth's atmosphere, they generate a problem too complicated to be analyzed by direct computation.

Such cases require the development of what Winsberg calls a *"discretized model equation."* This model is an idealized representation of the original model and usually has the following characteristics: First, continuous functions in the original model are represented with discrete analogues. This is nearly always done using numerical approximation techniques, but in large-scale simulations, theorists speak of the model's *grid*. This grid corresponds to the units of space that are basic to the model. The *resolution* of the model refers to the size of the grid. For example, the highest-resolution version of NOAA's Global Forecast System (GFS) weather model uses a horizontal grid roughly equal to 0.5 degrees of longitude and latitude and a 64-layer vertical grid whose elements increase in size as the atmosphere gets thinner (for details, see www.emc.ncep.noaa.gov).

The second component of the idealization is closely related to discretization and is called *parameterization*. This refers to the way the computed model represents phenomena that happen on scales smaller or faster than the grid size. Since the model must average over these effects within the grid, it has to represent this average at the resolution of the grid. These values are called parameters in the simulation literature, because they represent processes not captured in the model. Finally, an approximation scheme is devised which allows the now simplified, but still complex, functions to be computed in reasonable time frames. Such schemes are based on physically plausible assumptions about things such as processes that can be represented linearly that are not linear, interactions that can be reasonably ignored, and components of a model's intended scope that can be simply dropped from the analysis.

So there is a very wide range of techniques used for complete analysis of models. Even when the goal is complete analysis, it isn't always possible to achieve this goal. But there are also other possible goals of analysis, and in the next section I will discuss them.

5.2.2 Goal-Directed Analysis

Sometimes, models are constructed and analyzed to investigate a particular question. In such cases, analysis focuses on the features of the models relevant to that particular investigation. Many of the analytical techniques described above are also deployed in these cases, but with a narrow scope of investigation.

We can see this in a number of our examples. Let's begin with Volterra's analysis of his own family of models. Equations 2.3 and 2.4 are an uninstantiated model description. Different values set for the parameters r, m, a, and b will pick out different individual models. But the results of greatest interest to Volterra were the ones that held for the entire family of models (i.e., where the parameters are instantiated with different values), so he was able to carry out much of his analysis algebraically.

Given Volterra's goals, the most important property of his model was the one that ultimately allowed him to explain the anomalous Adriatic data. The *Volterra principle* states that, *ceteris paribus*, a general biocide will increase the relative abundance of the prey population. In Section 2.2 I have already discussed how Volterra was able to derive this conclusion. To recap, it involved finding the equilibrium values, which are also the time average for the two species, and expressing these as a ratio. This ratio has the form:

$$\rho = \frac{rb}{m} \tag{5.7}$$

Since general biocides will decrease the prey growth rate (r) and increase the predator death rate (m), $\rho(\text{biocide}) < \rho(\text{normal})$.

Another example of goal-directed analysis was the Army Corps' investigation of Reber's plan. I have already discussed in some detail how they calibrated the Bay model to its target (Section 2.1), so how did they show the Reber plan would have disastrous consequences? Strictly speaking, they did it by building a new model, one that was composed of the Bay model itself plus a set of structures that represented the Reber plan. The Corps describes the purpose of the test to "[d]etermine the effects of the barriers on the hydraulic and salinity regimes of the Bay on the downstream side of the barriers."

They continue:

> ... it was first necessary to establish hydraulic and salinity "base" tests which depicted the hydraulic and salinity characteristics throughout the model for existing conditions. Thus, a test in which no barrier was installed in the model is referred to as a "base" test, since its results constitute a basis of comparison for determining the effects of the barriers. Tests conducted with barriers installed in the model are designated "plan" tests. (Army Corps of Engineers, 1963b, 67)

In order to accomplish even the base test conditions, the model couldn't be simply switched on. The Bay system has fresh water coming in from the northeast and salt water coming in from the west. Part of the point of the model tests is to examine how the salinity gradient changes with the addition of barriers. So initially, the salt water had to be kept out of the northeastern part of the model until the salt- and freshwater flows were properly balanced.

> A salt-water gate across the San Pablo Strait was used in the model as an expedient to achieve stable hydraulic and salinity regimes with the least number of tidal cycles. Without the gate, water of ocean salinity would extend to the upper limit of the model before operation was started. Operation of the model for the base test for the 1956 tide was begun by flooding that portion of the model upstream from the salt-water gate with fresh water to near the elevation of lower low water; that portion of the model downstream of the gate was flooded with salt water (ocean salinity of 33 ppt) to the same elevation as that upstream from the gate. (67–68)

Now that the model was primed with salt water and fresh water in the proper locations, the gate was removed and the tide-control generators could be started. Salinity measurements were made periodically until the model began to approach the proper salinity conditions. As the model came close to the proper salinity conditions, the engineers prepared the freshwater inflows for the model.

> ... a simulated constant freshwater inflow into Suisun Bay of 16,000 cfs was started, the water being added to the surface of the flow in the model. The model was then operated an additional number of tidal cycles through an adjustment period until stable hydraulic and salinity regimens were obtained similar to that existing in the prototype on 21–22 September 1956. (68)

The Reber plan was studied by comparing measurements taken from this base test of the model, to a subsequent plan test. Here is how the engineers describe the base test:[2]

> When stable regimens had been obtained, complete hydraulic and salinity measurements were made throughout the Bay system. Tidal heights were measured with point gages at intervals corresponding to half-hourly intervals in the prototype for two complete tidal cycles (representing 49.6 hours in the prototype) at the 24 stations ... where tides had been observed in the prototype on 21–22 September 1956 during the field survey.
> ... Current velocities and directions were obtained at intervals corresponding to hourly intervals in the prototype for two complete tidal cycles at surface, mid-depth and bottom [...]. The velocity measurements were made in the model by miniature Price-type current meters; dye was used for determining current direction and precise time of slack water. In addition to salinity measurements with the conductivity cells, samples of water were drawn from the model at intervals corresponding to one-and-one-half to two-hour intervals in the prototype for one tidal cycle [...]. Salinity of the samples was determined chemically by titration with silver nitrate. (Army Corps of Engineers, 1963b, 68–69)

With the base tests complete, the Corps could compare these base saline and tidal measurements to the same measurements taken when the Reber plan was built within the model using sheet metal and plywood. In the initial studies the tidal heights, tidal phases, and velocities were compared between the base case and the Reber plan case.

The Reber Plan Barriers produced only slight effects in the elevations of HHW[3] and LLW,[4] and consequently the ranges of the tide at the various stations were changed

---

2. I will give a simplified explanation focusing only on the tidal target on 21–22 September 1956. In fact, the Corps actually examined two different tidal targets.
3. "High high water," in other words the higher of the two daily high tides.
4. "Low low water," in other words the lower of the two daily low tides.

only a small amount ... [and caused the tide] to occur from 25 minutes to one hour and 55 minutes earlier, except at Point Bonita [...].

The barriers reduced both flood and ebb current velocities oceanward of the barriers, including those in the Golden Gate at Station E, to values as low as zero to 0.1 ft/sec.

The Reber Plan Barriers changed the phases of the tidal currents, but since velocities within the Bay were reduced to 0.1 ft/sec, or less, there is no significance in phase change. (123–126)

The results quoted above begin to suggest the havoc the Reber Plan would have caused. The changes to water elevation and velocity are more significant than they might seem because of the scale of the Bay's tidal prism, which is the amount of water entering and leaving between low and high tides. The Bay's spring tidal prism is 2×10^9 cubic meters, or 528 billion gallons of water, and this amount of water flows out of the bay every 6.1 hours (Barnard, 2011). This creates a huge flushing effect of 24 million gallons per second at high tide, and these massive amounts of flowing water cause dilution and dispersion of pollutants. The Corps describes this situation as follows:

> The ability of estuarial waters to assimilate wastes discharged to it depends in no small measure upon their ability to disperse the wastes throughout large portions of their area by turbulent diffusion and tidal movement so as to bring about maximum dilution of the wastes. The speed with which the pollutants can be transferred out of the system to the ocean is likewise of significance. (Army Corps of Engineers, 1963b, 186)

Given this, the situation looked grim:

> A review of the effect of the [Reber] barrier on current velocities is helpful to an understanding of the effect of the barrier on dispersion and flushing characteristics of that part of the Bay still exposed to tidal action. In tests with the tide of 21–22 September 1956 ... the barrier reduced both flood and ebb current maximum velocities within the Bay oceanward of the barrier, including those in the Golden Gate, from the range of 2.0–4.8 ft/sec to the extremely low range of 0.0–0.1 ft/sec; and maximum velocities in the ocean over the Bar and in the Entrance Channel were reduced from the range of 1.8–2.9 ft/sec to the range of 0.2 to 1.0 ft/sec. (264)

These analyses indicate that the kinds of changes caused by the Reber plan's barriers might have had a very significant effect on the Bay's ability to flush out pollutants.

Since the model was at hand, the Corps put their concerns about flushing to the test. They designed a protocol whereby dye would be injected at specific points in the model, which corresponded to sewage outfalls as well as known areas of pollutant build-up. A full description of the analytical process is lengthy, so I will focus on a single dye release point which they designated 'I.' This point was located at the Passenger Pier Terminal in San Francisco, opposite North Point (264).

> Without the barrier, dye relased at "I" spread over a large part of the Bay system by HHW slack of the 1st cycle; to beyond San Francisco International Airport in South Bay, into the ocean near the outer limits of the Bar, through all of Central Bay, and two-thirds of the way across San Pablo Bay. ... With the barrier, not a trace of dye moved oceanward of the Golden Gate Bridge until between the 2nd and 3rd cycle; and by the 20th cycle the dye had progressed only half the distance to the outer limits of the Bar. Extremely large concentrations of dye persisted within a radius of several miles from the release point throughout the test.
>
> ... Extremely large concentrations of dye persisted throughout the tests, in the southern portion of Central Bay for release off of North Point, and in the northern portion of the Bay for release just downstream from the North Reber Barrier. The Reber Plan practically eliminated tidal current movements and associated turbulence, and thus changed the Bay oceanward of the barrier from a rapid to a slow dispersion and flushing system. (264–267)

The Corps thus warned that the Reber barriers would severely restrict the massive flushing action of the Bay caused by diurnal changes to the tidal prism. This would cause the build-up of domestic and industrial wastes. Col. John Kern, one of the directors of model studies in the post-Reber era, explained the situation this way:

> The Northern portion of our Bay is flushed by the inflow of the major rivers in Northern California (Sacramento and San Joaquin Rivers which drain the mountains and Central Valley). The southern portion of the Bay (San Francisco Bay) is very shallow and there is very little inflow from the few rivers and streams that drain into it. The only flushing that happens is that caused by the tidal changes. Pollutants such as sewage etc. which enter the southern Bay stay there for extended periods. (personal correspondence)

Moreover, the Corps' studies cast a skeptical light on Reber's claim that his plan would provide fresh water to the Bay Area.

> [The potential storage capacity of the Reber plan's lakes] is entirely negated by the possible and probable effective control of the runoff from the Sacramento-San Joaquin watershed for transfer flows. Local inflow into the Bay System, averaging around 550,000 acre-feet annually, would fail to meet the losses from evaporation, evapotranspiration, fish ladders and lockage, estimated to be 3,101,000 acre-feet annually under present conditions and to reach 4,732,00 by 2015. Should the effectual control of the delta inflow so throttle down the supply to a volume below the capability of local inflow to meet these losses, i.e., fall below a contribution of 2.6 million acre-feet at present and 4.2 million acre-feet by 2015, the lakes created

under the Reber Plan would shrink in a matter of several years to below mean sea level. For effective passage of navigation under such circumstances the lake levels would then necessarily have to be maintained by reversion to present conditions, i.e., opening the barriers for full access of the tidal prism, eliminating the salinity control and water conservation function for which the barriers would have been designed. (Army Corps of Engineers, 1963a, 184)

The Corps thus demonstrated with their model that the Reber plan would prevent the natural flushing action of the Bay, would waste more water than it could conserve, and would ultimately lead to massive and unanticipated changes to the Bay. This is, of course, to say nothing of the ecological consequences of the plan, which were only marginally considered by the Corps' studies. Most importantly for our purposes, this is an especially clear case of goal-directed analysis. Several specific aspects of a model were analyzed in order to answer specific research questions.

It is obviously a massive amount of hands-on work to study a concrete model such as the Bay model. For this reason, among others, theorists often choose to study computational models when they have specific information they are trying to discover. For example, imagine that a city planner was interested in learning whether there were any spatial configurations of houses in a Schelling-like model that would stabilize a racially integrated neighborhood. She could design a protocol whereby the Schelling model was initialized with different neighborhood configurations, and these were systematically searched for nonsegregated states.

Carrying out this search analytically is basically impossible. It is also extremely difficult and time-consuming to manually specify sets of initial conditions and then check each output. So how could a theorist carry out such an analysis? There are three important components: First, the theorist would need to specify the target behavior. In some cases, this would take the form of an *objective function* or *fitness function*, which she would aim to maximize. In this case, the relevant measure would be a particular type of model state: no clusters of like neighbors larger than a specified size. This would need to be further operationalized by specifying how many like neighbors within a certain neighborhood size each agent is allowed to have.

The second component of the search would require the specification of the search variables and parameters. Under what starting conditions would the model be examined? Would the population size vary, and if so, in what range? Would the utility functions vary? What range of distributions would be investigated and how would they be specified? The goal would be to know if we could achieve the relevant behavior under any of these combinations.

Finally, a method of searching this variable and parameter space would have to be chosen. The simplest, but most computationally intensive method would be a complete "sweep" of the variable and parameter values. This would mean trying every combination, and trying them repeatedly if the model contained probabilistic elements. There are also many other techniques for searching for the behavior of interest. One could simply do a random search, sampling from some probability distribution. If the behavior could be specified quantitatively, such as with an objective function, then one could use a hill-climbing technique, where behaviors close to the ones of interest are taken to be clues as to what combinations of values might yield the relevant behavior. Fancier methods would include the use of genetic algorithms, particle swarm optimization, or similar techniques. These methods all involve using information about a particular simulation of the model to choose the next parameter set, bringing the theorist closer to the solution she is looking for, assuming that it exists.

Since this chapter is about target-directed modeling, the simple case where we have one model and one target, the analytical methods I have discussed thus far made reference to a single model. But even in target-directed modeling, where a single model is ultimately fit to a target, a single model is rarely studied. For one thing, the kind of algebraic analysis I discussed above automatically gives theorists information about a family of models. In my example of exponential growth, every instantiation of the parameter r yields a distinct model. So any algebraic analysis applies to the entire family of exponential growth models.

But there are also reasons for going beyond the family of models generated from one uninstantiated model description. Sometimes, the key to understanding how particular parts of the model depend on other parts is by a method of *perturbations*. In some contexts, formal perturbation theory can be applied to a model in order to give an approximate, algebraic solution to the model (Bender & Orszag, 1999; Nayfeh, 2000). This technique proceeds by making a simpler version of the model—a model of the model if you like. A model description for this simpler model or family of models is written down and solved exactly. Then some of the simplifications are removed, generating solutions to a new model, which is still approximate, but closer to the original model. This process can be repeated, bringing the theorist closer and closer to the original model of interest.

Less formally, models can be perturbed by adding or subtracting components and one can study the new models that result from these changes. There are a number of reasons for this type of practice, but the one relevant to target-directed modeling is that such an analysis helps to identify how one part of the model depends on another. For example, some of the modern research about predation and the Volterra principle has investigated whether this principle continues to hold in closely related but distinct models. It can be shown that while many properties of the Lotka–Volterra model change when additional causal factors such as density-dependent growth, predator satiation, spatial structure, and prey-seeking cover are added, the Volterra principle occurs in each model. One conclusion is that the Voltera principle is a consequence of the kind of coupling associated with the predation (Weisberg & Reisman, 2008; see also Levins, 1962; Puccia & Levins, 1985; Justus, 2006).

Stepping back, we can see that the analysis aspect of target-directed modeling is a complex affair. Whether the goal is to have a complete understanding of the model, or just to learn about specific properties, analysis is an active process. The model must be manipulated either physically, mathematically, or computationally. While the exact manipulations that are required depend on the type of model, theorists use tests that can show how the parts of the model give rise to the model's behavior, and how changes to the model will generate changes in this behavior. Even when working with a model that has been calibrated with respect to a specific target, this aspect of target-directed modeling is done independently of the target.

5.3 ■ MODEL–TARGET COMPARISON

Although there are episodes of modeling that stop after the analysis stage (Section 7.2), most of the time theorists construct and analyze models as surrogates for target systems. In target-directed modeling, the target is a single, real system. Hence target-directed modeling requires a third stage in which the theorist attempts to coordinate a model with a real-world phenomenon.

In this section, and in my discussion of the model–world relation itself, I am going to focus on target systems and the relationship between targets and models. This is a different focus than most of the literature about model–world relationships because I will not primarily be focusing on experimentation, data, or confirmation. Before one can develop a theory of confirmation for models, one needs to be clearer about the nature of the model–world relationship. So in this section and in Chapter 8, I will be discussing model–world relations, not data analysis and confirmation.

5.3.1 Phenomena and Target Systems

Models are not compared directly to real phenomena, but to *target systems*, which are abstractions over these phenomena. When a scientist is interested in studying some phenomenon in the world, she begins by identifying a spatio-temporal region of interest. Perhaps she is interested in the breeding habits of Tasmanian devils over the last 100 years, or the height of the Mississippi River during the course of a year, or segregation in Philadelphia. The *phenomenon* of interest is the contents of this spatio-temporal region.

Phenomena have myriad properties, both static and dynamic. Call the entire set of these properties the *total state* of the phenomenon. In almost every instance, modelers are not interested in studying the total states of phenomena, but rather some scientifically important subset of these properties. These

Target-Directed Modeling ■ 91

restricted subsets are target systems. In other words, when scientists choose a focus, or an *intended scope* (see Section 3.3), they focus on some set of properties and abstract away the others. This yields a target system, a subset of the total state of the system.

For example, say that a biologist was interested in studying the remaining population of Tasmanian devils. The total state of the parts of Tasmania where the devils are still living would constitute the phenomenon of interest. She could then restrict her scope, abstracting to a number of different systems. For example, she might simply be interested in the devil population dynamics. Another possibility is that she is concerned about invasive species. In this case, she might abstract to a target composed of the carnivorous marsupials in the Tasmanian food web. Or maybe she is concerned about devil facial tumor disease and why it isn't recognized by the devils' immune system. This might require a target of devil population genetics. As this example shows, there are many possible target systems that can be generated from a single phenomenon. The relationship between phenomenon and target systems is thus one-to-(very)-many (see Figure 5.1).

The fact that the relationship between phenomenon and target systems is one-to-many opens the door for another possible proliferation. Since there are so many different targets that can be generated from the same phenomenon, does anything go? Are there standards that govern the kinds of abstractions that theorists make?

Answering this question is not straightforward. On the one hand, the one-to-many relationship of phenomena to targets is a desirable feature of theoretical practice. Choosing a target is very much in line with choosing a subject matter. Two scientists studying the Adriatic Sea after World War I might have very different interests: one in predator–prey relations, and one in the effect of surface temperature on algae blooms. The choice of one topic or the other will result

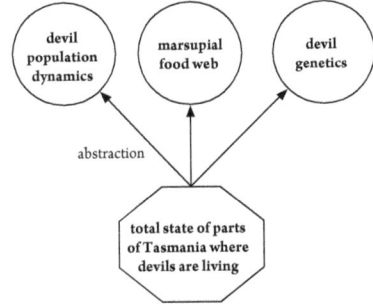

Figure 5.1 Left: a Tasmanian devil in captivity; photograph taken by the author. Right: a depiction of how abstracting away from a single phenomenon can yield many different kinds of targets.

in the generation of a different target system from the same phenomenon. The choice is free, but there seems to be no anarchy or arbitrariness in this difference.

Nevertheless, some further constraints are necessary to keep target generation rational. Abstraction comes in degrees, and there are many target systems that could be generated once a theorist decides to focus on predation or on algae blooms. Say she was interested in predation. Should the target include only population-level properties of the predators and prey? Should it include individual-level properties such as spatial locations? Should it include physiological information about satiation and digestion? Should it include abiotic factors of potential relevance such as clarity and temperature of the water?

Which particular features should be included in the construction of the target set is a subject for scientific research itself. When a scientific discipline is young, it is often very difficult to determine exactly which sets of features define fruitful research targets. This requires a trial and error approach, ultimately yielding the kinds of targets that make for the most fruitful research. Sometimes the choice of such features remains controversial, and may ultimately split a research community into subfields which cannot see eye to eye.

For example, ecological modeling generally falls into different theoretical camps. Population ecology studies the dynamics of population sizes, focusing primarily on phenomena such as competition, predation, growth, and mutualism. Community ecology looks at the interactions between populations and how these populations draw on biotic and abiotic resources in their environments. Even when studying the same phenomenon, say the Adriatic Sea, these subfields will abstract in different ways, including different properties in their targets (Elliott-Graves, 2012).

Thus, there can be considerable diversity in how different scientists construct targets from phenomena. However, there do seem to be some general principles that guide inclusion of features in targets. One principle is to include the properties of primary interest, and every factor that is causally linked to the properties of primary interest. Which factors this amounts to, of course, is a research topic. But theorists can still refer to a target, when its full extension is not known. Another possibility is for theorists to specify some limit of influence. For example, in studying the phenomenon of the Earth's movement around the Sun, theorists might restrict their space of targets so that only bodies which are influenced by the motion by some threshold would be included. So a target might include only the Sun and the Earth, or might add the moon and other nearby planets.

While a theorist can specify something about these goals in advance (Section 6.2), target selection is something that is often fine-tuned during the course of modeling. The procedure often looks like this: First, a target is chosen. Then, the model is compared to the target. Depending on the result of the comparison, the target can be fine-tuned, the model modified, or both.

The practice of modeling itself teaches scientists how to generate targets from phenomena.

5.3.2 Establishing the Fit between Model and Target

Once the theorist has a target system in hand, she can begin to establish the fit, or lack of fit, between the model and the target. Chapter 8 will be devoted to a general characterization of the model–world relation. In it, I will defend the view that models stand in a special kind of similarity relationship with their targets. However, it is possible to describe many aspects of the third stage of model building independently of any particular account of this model–target relationship, and that is what I will do in the remainder of this chapter. In this section, I will simply refer to the model–world relation as a matter of *fit*. A model can be successfully applied to a target when it fits the target.

As I said above, fitting models to the world does not depend on the total states of phenomena, but rather on abstracted target systems. Moreover, model–target fits do not necessarily put equal weight on all aspects of the model and target, nor are they uniform in the degree of fit that must be established between each property of the model and of the target. Modelers' fidelity criteria specify which properties must fit and to what degree they must fit (see Sections 3.3 and 8.5).

For example, imagine that a modeler wanted to assess the extent to which the Lotka–Volterra model could be used to explain the population dynamics of Canadian lynxes and hares, a predator–prey system that has been carefully studied over many years (Elton & Nicholson, 1942; Moran, 1953; May, 1972; Gilpin, 1973). A very simple kind of assessment of fit is to adjust the independent variables and parameters of the Lotka–Volterra equation to find the best-fitting model from the model family generated by this equation. The best-fitting model can then be plotted against the data and the goodness of fit can be assessed. This is depicted in Figure 5.2.

As one can see from the plot depicted in Figure 5.2, the Lotka–Volterra model is definitely not a perfect fit to the data. Moreover, there is reason to doubt that the oscillating hare population is the primary cause of the lynx oscillation and vice versa, based on analysis of the fit and on field observations. So even this simple fitting of the model's output to a representation of the target is clearly a nontrivial matter (Stenseth, Falck, Bjørnstad, & Krebs, 1997).

This example also indicates another distinction between kinds of model–target fit, which sometimes goes by the name of *calibration*. Target-directed modeling really covers two different kinds of model practices. In one, the goal of the model is to get a qualitative fit between model and target. The idea is to see if we can more or less reproduce the phenomenon of interest using a model. This is an *uncalibrated* use of a model.

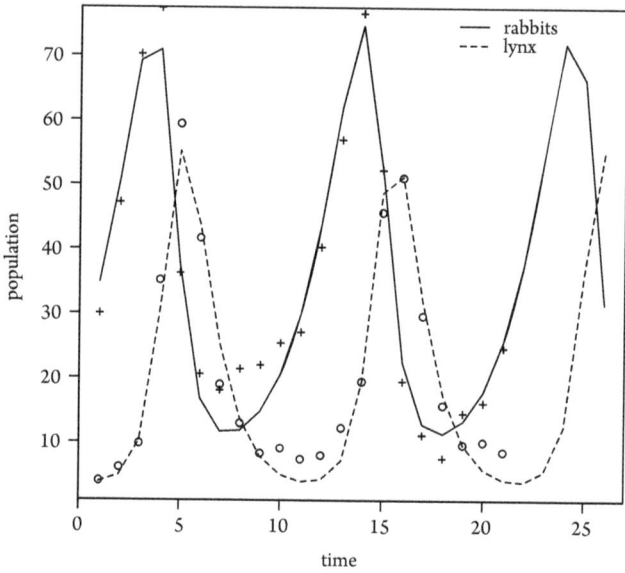

Figure 5.2 A set of observations for Canadian lynx population sizes fit to the Lotka–Volterra model.

One way to think about Schelling's segregation model is as an uncalibrated model of segregation in an actual city. Imagine that the central dynamic of this model, moving away because you don't have enough like neighbors, was the primary explanation of segregation in a city like Philadelphia. If this were the case, we would never expect more than a qualitative, uncalibrated fit of the model to the city. There simply isn't enough detail in the model to correctly reproduce some of the finer details such as cluster size, cluster extent, cluster position, time course, and so on. What the model shows is that a particular utility function and distribution of people can cause segregation.

We could look at such uncalibrated uses of a model as being calibrated, but to a very abstract target. While nothing precludes this, the level of abstraction would need to be very high indeed. If the target was "Philadelphia has segregation," as opposed to any details about the segregation, we could almost certainly achieve an exact fit between the model and the target. But it seems very unlikely that this would constitute a target of much interest to scientists.

In *calibrated* modeling, the goal is to make an extremely high-fidelity model of a target. Depending on the target, such cases may require considerably more complex models. But they also may require considerably more work to calibrate. Consider the San Francisco Bay model. Even after the model was constructed, it took the Army Corps 18 months to calibrate the model to its target, which was the average tidal cycle for 36 hours of 21–22 September 1956. Calibration involved making small adjustments to the model such as placing pieces of copper strips on the bottom in order to correctly generate the tides. This required a feedback process: adjustment of the model, then analysis of the model, then further modifications to the model until the model reached a specified standard of fidelity.

What we have learned in this section is that establishing the fit of the model to a phenomenon in the world happens in two distinct stages. First, the theorist must subject the phenomenon to a process of abstraction, deciding what aspects of the phenomenon she wishes to study. This yields the target. With target in hand, the theorist can proceed to try to fit a calibrated or uncalibrated model to the target at the desired level of fidelity. The interaction between fit, fidelity, and target will be discussed in more detail in Chapter 8.

5.3.3 Representations of Targets

Before closing this section, I want to consider a potential problem for this account. Target systems are simply abstractions over phenomena, subsets of total states of systems. The properties that compose these states are concrete properties. In some kinds of concrete modeling, the target can be directly compared to a model. But how can mathematical and computational models be compared to concrete targets? What possible similarities do they have?

I think this objection can easily be resolved. Mathematical and computational models, as well as concrete models in some cases, are compared to mathematical representations of targets,[5] not the targets themselves. Each state of the target is mapped to some mathematical space. In simple, dynamical models, the mapping is such that the major determinable properties (e.g., species abundance, pressure, time, temperature, etc.) of the target are mapped to dimensions of a state space, and specific states are mapped to points in this space. Now one interpreted mathematical object can be compared to another, and we avoid problems about comparing mathematical properties to concrete properties.[6]

If this procedure sounds familiar, it is because it is nearly identical to the way I discussed the construction of mathematical models out of mathematical structures. Mathematical representations of targets are the same kinds of structures as mathematical models. The difference is that the structure of a mathematical model is a free choice of the theorist, but the mathematical representation of the target is a highly constrained representation of the way that part of the world is.

---

5. In Weisberg, 2003, I called this a "parameterized target system," but no longer use this term because "to parameterize" means something different in the simulation literature.
6. The construction of such a mathematical or computational representation of a target is very similar to what Suppes called a *model of data* (1960b).

Targets and their mathematical representations are not the same as empirically collected data. Targets are sets of real-world properties and mathematical representations of targets are representations of these properties. Measurements and observations, along with background theory, statistics, and computation, are used to make inferences about the nature of targets and mathematical representations of targets. These inference steps are far from trivial. However, the details about how they work are part of the literature on experimentation and confirmation theory and not my main focus here. In practice, of course, all theorists can do is try to compare their models to their best current picture about the nature of targets. Assessments about goodness of fit are with respect to their current best knowledge.

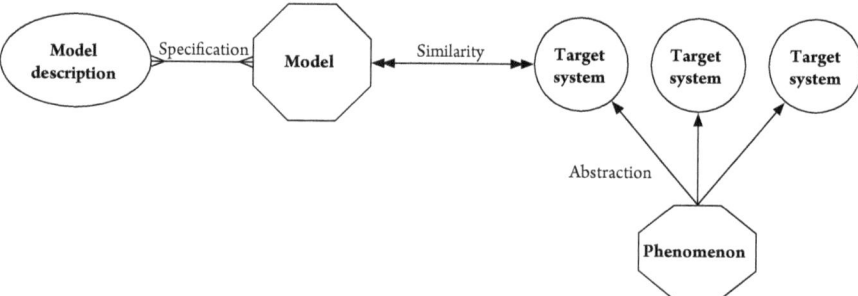

Figure 5.3 The relationship between phenomenon, target, and model for a concrete model.

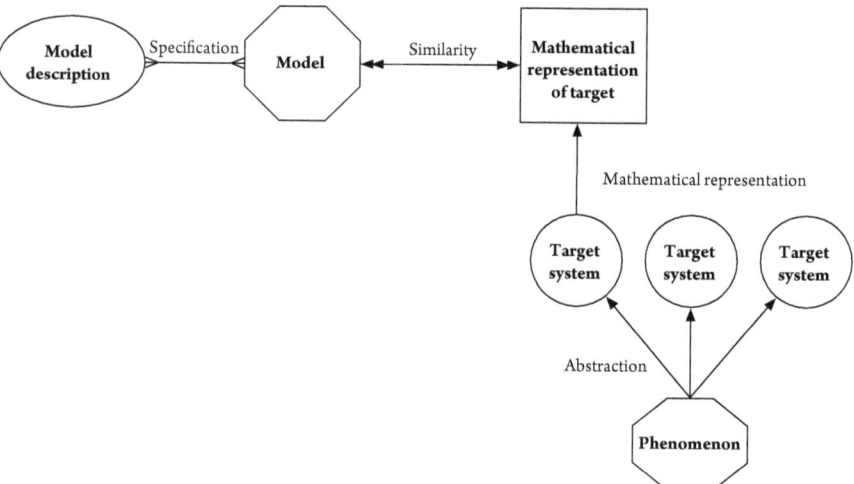

Figure 5.4 The relationship between phenomenon, target, mathematical representation of target, and model for a mathematical model.

In this chapter, I have argued that there are three aspects to target-directed modeling: developing the model, analyzing the model, and fitting the model to a target. When the theorist completes these three tasks, or takes them on board from someone else, she is in a position to do model-based science. She can learn about a real-world target by studying a model. This picture about the relationships between models and their targets is shown for concrete models in Figure 5.3 and for mathematical and computational models in Figure 5.4.

As I said at the beginning of the chapter, target-directed modeling is the simplest kind of modeling, in which there is a single model and a single target. The next two chapters will investigate more complex kinds of modeling. Since these complexities are often generated by *idealization*, the intentional distortion of aspects of a target that are represented in a model, I turn to this concept in the next chapter.

6 Idealization

Target-directed modeling aims to give a realistic picture of a single target using a single model. In this chapter and the next, I am going to discuss more complex versions of modeling. In this chapter, I will talk about idealization, where the model isn't maximally realistic, or where multiple, partial models for a single target are generated and analyzed.

Philosophers of science increasingly recognize the importance of idealization, which is the intentional introduction of distortion into scientific representations. This is especially true in the literature about models and modeling, led in no small part by the pioneering work of Nowak (1972), Cartwright (1983), McMullin (1985), Wimsatt (1987), and Giere (1988).

According to a straightforward view, we can think of idealization as a departure from complete, veridical representation of real-world phenomena. In idealization, we *distort* or misrepresent our target by representing it as having properties that it does not have (Jones, 2005). In other words, a model is idealized with respect to its target when it fails to represent some important aspects of the target.

The literature about idealization has presented a number of distinct accounts of what this kind of distortion amounts to. I believe that this diversity of accounts reflects the fact that there isn't a single way to generate a distorted representation of a target. So rather than choosing and developing a single account of idealization, I will regard the extant literature as giving evidence that there are three distinct forms of idealization.

6.1 ■ THREE KINDS OF IDEALIZATION

One of the most important insights of the modern idealization literature is that idealization should be seen as an activity that involves distorting models, not simply as a property of the theory–world relationship. This suggests that in order to study idealization we will need to know what activity is characteristic of that form of idealization and how that activity is justified. These activities and justifications can be grouped into three kinds of idealization: *Galilean idealization, minimalist idealization,* and *multiple-models idealization*.

6.1.1 Galilean Idealization

Galilean idealization is the practice of introducing distortions into models with the goal of simplifying, in order to make them more mathematically or computationally tractable. One starts with some idea of what a nonidealized representation of the target would look like. Then one creates a model of the target that is simplified and distorted with respect to the nonidealized representation. Galilean idealization was most thoroughly characterized by McMullin who sees the point of this kind of idealization as "grasp[ing] the real world from which the idealization takes its origin" (1985, 248) by making the problem simpler, and hence more tractable.

This technique is named after Galileo, because he employed it in both theoretical and experimental investigations. Although this book is concerned with modeling, and hence with theoretical investigation, Galileo's vivid description of the experimental version is useful for conceptualizing the basic notion of Galilean idealization. When discussing the determination of gravitational acceleration in the absence of a medium devoid of resistance, Galileo suggests a kind of experimental idealization:

> We are trying to investigate what would happen to moveables very diverse in weight, in a medium quite devoid of resistance, so that the whole difference of speed existing between these moveables would have to be referred to inequality of weight alone.... Since we lack such a space, let us (instead) observe what happens in the thinnest and least resistant media, comparing this with what happens in others less thin and more resistant. (quoted in McMullin, 1985, 267)

Lacking a medium devoid of resistance, Galileo suggests that we can make some progress on the problem by initially using an experimental setup similar to the envisioned situation. After understanding this system, the scientist systematically removes the effect of the introduced distortion. The same type of procedure can be carried out in theorizing: introduction of distortion to make a problem more tractable, then systematic removal of the distorting factors.

Galilean idealization is justified pragmatically. We simplify to more computationally tractable theories in order to be able to learn from them. If the theorist had not idealized, she would have been in a worse situation, stuck with an intractable model. Since the justification is pragmatic and tied to tractability, advances in computational power and mathematical techniques should lead the Galilean idealizer to de-idealize, removing distortion and adding back detail to her theories. With such advances, McMullin argues, "models can be made more specific by eliminating simplifying assumptions and 'de-idealization', as it were. The model then serves as the basis for a continuing research program" (261). Thus the justification and rationale of Galilean idealization is not only pragmatic, it is highly sensitive to the current state of a particular science.

Galilean idealization is important in research traditions dealing with computationally complex systems. Computational chemists, for example, calculate molecular properties by computing approximate wavefunctions for molecules of interest. At first, all but the simplest systems were intractable. When electronic computers were introduced to computational chemistry, calculated wavefunctions remained crude approximations, but more complex, chemically interesting systems could be handled.

As computational power has continued to increase in the 21st century, it has become possible to compute extremely accurate (but still approximate) wavefunctions for moderate-sized molecules. Theorists in this tradition aim to develop ever better approximations for molecular systems of even greater complexity. These techniques are still approximate, but research continues to bring computational chemists closer to the goal of "[calculating] the exact solution to the Schrödinger equation, the limit toward which all approximate methods strive" (Foresman & Frisch, 1996, 95).

This example nicely summarizes the key features of Galilean idealization. The practice is largely pragmatic; theorists idealize for reasons of computational tractability. The practice is also nonpermanent. Galilean idealization takes place with the expectation of future de-idealization and more accurate representation.

6.1.2 Minimalist Idealization

Minimalist idealization is the practice of constructing and studying models that include only the core causal factors which give rise to a phenomenon. These are often called *minimal models* of the phenomenon. Put more explicitly, a minimal model contains only those factors that make a difference to the occurrence and essential character of the phenomenon in question.

A classic example of a minimalist model in the physical sciences is the one-dimensional Ising model. This simple model represents atoms, molecules, or other particles as points along a line and allows these points to be in one of two states. Originally, Ernst Ising developed this model to investigate the ferromagnetic properties of metals. It was further developed and extended to study many other phenomena of interest involving phase changes and critical phenomena.

The model is powerful and allows qualitative and some quantitative properties of substances to be determined. But it is extremely simple, having almost no realistic detail about the substances being modeled. What it seems to capture are the interactions and structures that really make a difference to the occurrence of the phenomenon. In other words, it captures the core causal factors giving rise to the target.

Among recent discussions of idealization in the philosophical literature, minimalist idealization has been the most comprehensively explored position. As such, there is some diversity among the articulations of this position. One view is Michael Strevens' *kairetic* account of scientific explanation (Strevens, 2004, 2008). This account of explanation is causal; to explain a phenomenon is to give a causal story about why that phenomenon occurred. What makes Strevens' account distinct is that the explanatory causal story is limited to only those factors that made a difference to the occurrence of the phenomenon.

"Making a difference" is a fairly intuitive notion, but Strevens defines it explicitly in terms of what he calls *causal entailment*, which is logical entailment in a causal model. The causal structure is first represented propositionally in a model. Logical entailments within this model represent causal entailments in the target. A causal factor makes a difference to a phenomenon just in case its removal from a causal model prevents the model from entailing the phenomenon's occurrence. A causal model of the difference-making factors alone is called a *canonical* explanation of the target phenomenon.

For Strevens, idealization is the introduction of false but nondifference-making causal factors to a canonical explanation. In explaining Boyle's law, for example, theorists often introduce the assumption that gas molecules do not collide with each other. This assumption is false; collisions do occur in low-pressure gases. However, low-pressure gases behave as if there were no collisions. This means that collisions make no difference to the phenomenon and are not included in the canonical explanation. Theorists' explicit introduction of the no-collision assumption is a way of asserting that collisions are actually irrelevant and make no difference. Even with this added, irrelevant factor, the model is still minimalist because it accurately captures the core causal factors.

Other accounts of minimalist idealization associate minimalism with generation of the canonical explanation alone. Robert Batterman's account of asymptotic explanation is an example of such a view. Asymptotic methods are used by physicists to study the behavior of model systems at the limits of certain physical magnitudes. These methods allow theorists to study how systems would behave if certain effects were removed, which allows the construction of "highly idealized minimal models of the universal, repeatable features of a system" (Batterman, 2002, 36; see also Batterman, 2001). These minimal models have a special role in physics because they can be used to explain universal patterns, common behaviors across material domains such as pressure, temperature, and critical phenomena. Adding more detail to the minimal model does not improve the explanations of these patterns; more details only allow a more thorough characterization of a highly specific event.

Arguing in a similar vein, Stephan Hartmann describes cases where complex systems are characterized using physical models "of (simple) effective degrees of freedom," which help to give us "partial understanding of the relevant mechanisms for the process under study." This plays a cognitive role by allowing theorists "to get some insight into the highly complicated dynamics" of such systems (Hartmann, 1998, 118).

Nancy Cartwright's account of abstraction is also an example of what I call minimalist idealization. On her view, abstraction is a mental operation, where we "strip away—in our imagination—all that is irrelevant to the concerns of the moment to focus on some single property or set of properties, 'as if they were separate'" (Cartwright, 1989, 187). If the theorist makes a mathematical model of this abstract, real phenomenon, then she is in possession of a minimal model. Such a model can reveal the most important causal powers at the heart of a phenomenon.

Despite the differences between minimalist idealization and Galilean idealization, minimalist idealizers could in principle produce an identical model to Galilean idealizers. For example, imagine that we wanted to model the vibrational properties of a covalent bond. A standard way to do this is to use a harmonic oscillator model. This model treats the vibrating bond as spring-like with a natural vibrational frequency due to a restoring force. This is a very simple representation of the vibrational properties of a covalent bond, but one that is commonly used in spectroscopy. Galilean idealizers would justify the use of this model by saying that it is pragmatically useful for calculating energies, thus avoiding having to calculate the many-dimensional potential-energy surface for the whole molecule. Minimalist idealizers, however, would justify the use of this model by suggesting that it captures what really matters about the vibrations of covalent bonds. The extra detail in the full potential-energy surface, they would argue, is extraneous.

As this example illustrates, the most important differences between Galilean and minimalist idealization are the ways that they are justified. Even when they produce the same representations, they can be distinguished by the rationales they give for idealization. Further, while Galilean idealization ought to abate as science progresses, this is not the case for minimalist idealization. Progress in science and increases in computational power should drive the two apart, even if they generate the same model at a particular time.

Just as there is no single account of minimalist idealization, there is no single account of its justification. However, all of the influential accounts described above agree that minimalist idealization should be justified with respect to the cognitive role of minimal models: They aid in scientific explanations. Hartmann argues that minimal models literally tell us how phenomena behave in a simpler world than our own. This gives us the necessary information to explain real-world phenomena. For Batterman, minimal models demonstrate how fundamental structural properties of a system generate common patterns among disparate phenomena. Strevens and Cartwright look at things more causally, describing the role of minimal models as showing us the causal factors that bring about the phenomenon of interest.

In all of these cases, minimalist idealization is connected to scientific explanation. Minimal models isolate the explanatorily causal factors either directly (Cartwright and Strevens), asymptotically (Batterman), or via counterfactual reasoning (Hartmann). In each case, the key to explanation is a special set of explanatorily privileged causal factors. Minimalist idealization is what isolates these causes and thus plays a crucial role for explanation. This means that, unlike Galilean idealization, minimalist idealization is not at all pragmatic and we should not expect it to abate with the progress of science.

6.1.3 Multiple-Models Idealization

Multiple-models idealization (hereafter MMI) is the practice of building multiple related but incompatible models, each of which makes distinct claims about the nature and causal structure giving rise to a phenomenon. MMI is similar to minimalist idealization in that it is not justified by the possibility of de-idealization back to the full representation. However, it differs from both Galilean and minimalist idealization in not expecting a single best model to be generated.

One most commonly encounters MMI in sciences dealing with highly complex phenomena. In ecology, for example, one finds theorists constructing multiple models of phenomena such as predation, each of which contains different idealizing assumptions, approximations, and simplifications. Chemists continue to rely on both the molecular orbital and valence bond models of chemical bonding, which make different, incompatible assumptions. In a dramatic example of MMI, the United States National Weather Service (NWS) employs several different models of global circulation patterns to model the weather. Each of these models contains different idealizing assumptions about the basic physical processes involved in weather formation. Although attempts have been made to build a single model of global weather, the NWS has determined that the best way to make high-fidelity predictions is to employ all three models, despite the considerable expense of doing so.

The literature about MMI is less well developed than the literature about the others, so there is less of a clear consensus about its justification. But one especially important justification of MMI is the existence of tradeoffs, a position closely associated with biologist Richard Levins and his philosophical allies. This justification begins by noting that theorists have different goals for their representations, such as accuracy, precision, generality and simplicity (Levins, 1966; Weisberg, 2006; Matthewson & Weisberg, 2009).

Levins further argues that these desiderata and others can trade off with one another in certain circumstances, meaning that no single model can have all of these properties to the highest magnitude. If a theorist wants to achieve high degrees of generality, accuracy, precision, and simplicity, she will need to construct multiple models. Levins summarizes his discussion of these issues as follows:

> The multiplicity of models is imposed by the contradictory demands of a complex, heterogeneous nature and a mind that can only cope with few variables at a time; by the contradictory desiderata of generality, realism, and precision; by the need to understand and also to control; even by the opposing esthetic standards which emphasize the stark simplicity and power of a general theorem as against the richness and the diversity of living nature. These conflicts are irreconcilable. Therefore, the alternative approaches even of contending schools are part of a larger mixed strategy. But the conflict is about method, not nature, for the individual models, while they are essential for understanding reality, should not be confused with that reality itself. (Levins, 1966, 27)

Our cognitive limitations, the complexity of the world, and constraints imposed by logic, mathematics, and the nature of representation, conspire against simultaneously achieving all of our scientific desiderata. Thus, according to Levins, communities of scientists should construct multiple models, which collectively can satisfy our scientific needs.

Several other justifications for MMI can be found in the literature. A particularly important one is Wimsatt's (1987) observation that highly idealized models taken together help us develop truer theories. Similarly, population biologists Joan Roughgarden (1979) and Robert May (2001, 2004) argue that clusters of simple models increase the generality of a theoretical framework, which can lead to greater explanatory depth.

Some of these motivations suggest strong parallels between MMI and minimalist idealization. In some cases, one cannot build a single minimal model that contains all of the core causal factors for a class of phenomena. Yet it may be possible, in such cases, to build a small set of models, each of which highlights a different factor and which together account for all of the core causal factors. This motivation for MMI is parallel to the motivation for minimalist idealization, even though the practice itself is different.

However, there are additional motivations for engaging in MMI that do not parallel the motivation for minimalist idealization. For example, modelers may engage in MMI strictly for the purpose of maximizing predictive power, as do the forecasters at the National Weather Service. Another instance of MMI may involve building a set of models that gives maximum generality, at the expense of capturing all of the core causal factors. Still another is the synthetic chemist's or engineer's motivation for MMI: to find the set of idealized models that is maximally useful for creating new structures. There are thus many motivations for MMI. Some are pragmatic, where scientists are focused on prediction and structure construction, while some are explanatory and nonpragmatic.

MMI also gives a complex, mixed answer about the permanence of idealization as science progresses. In some domains, MMI may abate with the progress of science. For example, the National Weather Service may one day discover a single model that makes optimal predictions. However, if tradeoffs exist between theoretically important desiderata in a particular domain, then we should not expect MMI to abate with further progress. These tradeoffs are consequences of logic and mathematics and thus present a permanent justification for MMI.

From the discussion so far, it may seem that the literature on idealization describes a hodgepodge of disparate practices, leaving no hope for any further analysis of idealization simpliciter. This worry is not without merit because the methods, goals, and justifications of these three forms of idealization are quite distinct. Although a fully unified account of the three kinds of idealization is impossible, some progress can be made towards developing a unified framework with which to understand the practice of idealization in general. This framework focuses on the goals associated with idealization, rather than the activities or products of it. I will call these goals the *representational ideals* of idealization, and will argue that they are encoded as part of theorists' construals.

6.2 ■ REPRESENTATIONAL IDEALS AND FIDELITY CRITERIA

Representational ideals are the goals governing the construction, analysis, and evaluation of theoretical models. They regulate which factors are to be included in models, set up the standards theorists use to evaluate their models, and guide the direction of theoretical inquiry. Representational ideals can be thought of as having two components: inclusion rules and fidelity rules. In other words, they specify constraints on a theorist's intended scope and fidelity criteria (see Section 3.3). Inclusion rules tell the theorist which kinds of properties of the phenomenon of interest, or target system, must be included in the model, while fidelity rules concern the degrees of precision and accuracy with which each part of the model is to be judged.

An important, albeit very simple, representational ideal is called COMPLETENESS, which is associated with classic accounts of scientific method. I will discuss it first because it forms an important background against which every kind of idealization can be discussed.

6.2.1 COMPLETENESS

According to COMPLETENESS, the best theoretical description of a phenomenon is a complete representation. The relevant sense of "completeness" has two components associated with it: inclusion rules and fidelity rules, respectively. The inclusion rules state that each property of the target phenomenon must be included in the model. Additionally, anything external to the phenomenon that gives rise to its properties must also be included in the model. Finally, structural and causal relationships within the target phenomenon must be reflected in the structure of the model. COMPLETENESS fidelity rules tell the theorist that the best model is one that represents every aspect of the target system and its exogenous causes with an arbitrarily high degree of precision and accuracy.

The description of COMPLETENESS given so far is accurate, but potentially misleading. With very few exceptions, the inclusion and fidelity rules of COMPLETENESS set a goal that is impossible to achieve. Unless extremely self-deceived, or in possession of an extremely simple and abstract target system, no theorist thinks that complete representation is actually possible. Given the impossibility of achieving complete representation, how can COMPLETENESS play a guiding role in scientific inquiry?

Despite its unattainable demands, COMPLETENESS can guide inquiry in two ways. First, COMPLETENESS sets up a scale with which one can evaluate all representations, including suboptimal ones. If a theorist wants to rank several representations of the same phenomenon and has adopted COMPLETENESS, she has a straightforward way to do so. The closer a representation comes to completeness, the better it scores. I call this the evaluative function of the representational ideal because it sets the standards for evaluating suboptimal representations.

The second and more important way that COMPLETENESS can guide inquiry is through its regulative function. Regulative functions are similar to what Kant called *regulative ideals* (Kant, 1998, A642/B670). They do not describe a cognitive achievement that is literally possible; rather, they describe a target or aim point. They give the theorist guidance about what she should strive for and the proper direction for the advancement of her research program. If a theorist adopts COMPLETENESS, she knows that she should always strive to add more detail, more complexity, and more precision to her models. This will bring her closer to the ideal of completeness, although in most cases, she will never fully realize this goal.

COMPLETENESS is a unique representational ideal because it directs theorists to include everything in their representations. All other ideals will build in some aspect of approximation or distortion. In thinking about ideals other than COMPLETENESS, we can begin to see the outline of a framework for characterizing the three kinds of idealization. Different kinds of idealization will be associated with different representational ideals. Before we carry this analysis forward, let us consider several additional representational ideals.

6.2.2 SIMPLICITY

After COMPLETENESS, the next most straightforward ideal is SIMPLICITY. The inclusion rule for this ideal councils the theorist to include as little as possible, while still being consistent with the fidelity rules. The fidelity rules for SIMPLICITY demand a qualitative match between the behavior of target system and the properties and dynamics of the model.

SIMPLICITY is primarily employed by working scientists in two contexts. The first is pedagogical. Students are often introduced to the simplest possible model that can make sense of the data, even where scientists believe that the model contains serious distortions. One example of this is the use of G. N. Lewis' electron-pair model of chemical bonding (G. N. Lewis, 1916). In this model, chemical bonds are treated as electron pairs shared between two atoms. Although Lewis' model was developed before quantum mechanics, which gives our best understanding of chemical bonds, it is a surprisingly useful heuristic for predicting the structure of many molecules, especially small ones. Beginning students are thus introduced to this model as a way of building intuitions about chemical structure and reactivity. Working scientists, however, would never take this model to give an accurate accounting of a molecule's electronic structure.

The second scientific context where SIMPLICITY is employed is when theorists construct models to test general ideas. "A minimal model for an idea tries to illuminate a hypothesis.... [It] is not intended to be tested literally, any more than one would test whether the models for a frictionless pulley or a frictionless inclined plane are wrong" (Roughgarden, 1997, x). This second use represents a motivation and justification for a particular kind of modeling in scientific practice. Theorists often begin a project by trying to determine what kind of minimal structures could generate a property of interest. They do not need to know, at first, how a specific target system actually works. Once the dynamics are understood in simple models, theorists examine more complex models and empirical data to assess the plausibility of the simple model's explanation of a real system's behavior.

6.2.3 1-CAUSAL

Another representational ideal called 1-CAUSAL instructs the theorist to include in the model only the core or primary causal factors that give rise to the phenomenon of interest. Put in the language of the causation literature, this ideal tells the theorist to include only the factors that make a difference. The theorist constructs a model of a much simpler system than the one actually being studied, one that excludes higher-order causal factors. Higher-order causal factors are those that make no difference to the occurrence of the phenomenon, but control the precise way in which the phenomenon occurs.

This representational ideal is closely related to SIMPLICITY, but unlike SIMPLICITY, 1-CAUSAL restricts the level of simplicity that is allowed. If we are trying to construct the simplest possible model that can make predictions qualitatively compatible with our observations, there is no restriction on the kind or number of causal factors that must be included. SIMPLICITY, for example, may allow us to neglect all quantum mechanical effects and use the Lewis model. 1-CAUSAL, however, would not sanction the use of such a model because it requires the theorist to include the quantum mechanical interactions that are actually responsible for a molecule's structure.

1-CAUSAL's fidelity criteria make a considerable difference in determining when the theorist has constructed an adequate model because its inclusion rule (restriction to primary causal factors) is not very specific. In addition, the fineness of specification of the target phenomenon itself will make a difference to the kind of model we can build. Imagine that we wanted to build a 1-CAUSAL model for the maintenance of the sex ratio. We would need a more complex model to explain the 1.05:1 ratio of male to female *Homo sapiens* than if we were interested only in why the sex ratio is roughly 1:1. Even holding the fidelity criteria fixed, the best model would be different in these two cases, with the former requiring greater specification of internal and external causal factors.

Models generated using 1-CAUSAL are especially useful in two contexts. Like the models generated with SIMPLICITY, they can be used as starting points for the formulation and analysis of more complex models. 1-CAUSAL models are typically generated when one has a reasonably comprehensive understanding of how a system behaves, since knowing the primary causal factors that give rise to a phenomenon requires knowing quite a lot about the system. Further modeling from this point is usually aimed at greater quantitative accuracy, not deeper fundamental understanding.

The second context where 1-CAUSAL is especially important involves scientific explanation. Several recent philosophical accounts of scientific explanation have pointed to the central role that primary causal factors—the factors that really make a difference—play in scientific explanation. Recent work on the cognitive psychology of explanation has also emphasized the crucial role that picking out central causal factors plays in people's judgments of explanatory goodness. In their methodological discussions, a number of other scientists have commented on this connection. For example, chemist Roald Hoffmann emphasizes that "if understanding is sought, simpler models, not necessarily the best and predicting all observables in detail, will have value. Such models may highlight important causes and channels" (Hoffmann, Minkin, & Carpenter, 1996). These accounts all suggest that models generated with 1-CAUSAL seem to be at the heart of theorists's explanatory practices.

6.2.4 MAXOUT

We now move from an ideal which looks superficially like SIMPLICITY to one that looks superficially like COMPLETENESS, the ideal called MAXOUT. This ideal says that the theorist should maximize the precision and accuracy of the model's output. It says nothing, however, about how this is to be accomplished.

One way to work towards this ideal is by constructing highly accurate models of every property and causal factor affecting the target. This is the same approach taken in COMPLETENESS, although the goal of MAXOUT is to achieve maximum output precision and accuracy, not a complete representation. A second option, one more commonly associated with MAXOUT, is to engage in model selection, a process of using statistics to choose a functional form, parameter set, and parameter values which best fit a large data set. The model selected by these techniques is then continually optimized as further data comes in. Finally, MAXOUT also sanctions the use of black-box models, the sort that have amazing predictive power, but for unknown reasons. These may be discovered using model selection techniques, or may be discovered in a more serendipitous fashion.

At first blush, it may seem unscientific to adopt an ideal that values predictive power over everything else. Most scientists believe that their inquiry is aimed at more than raw predictive power. While scientists want to know how a system will behave in the future, they also want an explanation of why it behaves the way that it does. MAXOUT ensures that we will generate models which are useful for predicting future states of the target system, but gives no guarantee that the models will be useful for explaining the behavior of the system.

Nevertheless, representations generated by MAXOUT have their place in scientific inquiry. Explanation and prediction are clearly both important goals of scientists, but there is no reason that they must both be fulfilled with the same model. Theorists can adopt a mixed representational strategy, using different kinds of models to achieve different scientific goals. It may also be rational to elevate predictive power above all other considerations in some situations. Following his reflection on the importance of simple models quoted above, Hoffmann argues that "If predictability is sought at all cost—and realities of marketplace and judgments of the future of humanity may demand this—then simplicity may be irrelevant" (Hoffmann, Minkin, & Carpenter, 1996).

6.2.5 P-GENERAL

Generality is a desideratum of many types of modeling. This desideratum really has two distinct parts: *a-generality* and *p-generality*. A-generality is the number of *actual* targets a particular model applies to given the theorists' adopted fidelity criteria. P-generality, however, is the number of *possible*, but not necessarily actual, targets a particular model captures (Matthewson & Weisberg, 2009). The representational ideal P-GENERAL says that considerations of p-generality should drive the construction and evaluation of theoretical models.

While a-generality may seem like the more important kind of generality, theorists are often interested in p-generality for several reasons. P-general models can be part of the most widely applicable theoretical frameworks, allowing real and nonreal target systems to be compared. P-generality is also often thought to be associated with explanatory power. This can be seen in both the philosophical literature on explanation and in the comments of theorists. An excellent example of this can be found in Arthur Eddington's oft-quoted dictum: "We need scarcely add that the contemplation in natural science of a wider domain than the actual leads to a far better understanding of the actual" (Eddington, 1927).

P-GENERAL can also play a subtler regulative role. Instead of trying to understand specific targets, theorists may wish to understand fundamental relationships or interactions, abstracted away from actual systems. For example, ecologists may wish to study predation or competition, far removed from the interactions of particular species. In such cases, P-GENERAL is often adopted, guiding theorists to develop models that can be applied to many real and possible targets. This exploratory activity is a very important part of modern theoretical practice, although we do not yet have a good philosophical account of how it works. One thing we do know, however, is that there is a delicate balance between achieving deep and insightful p-generality and low-fidelity, uninformative p-generality, generated by overly simplistic models.

We have now looked at a number of representational ideals, the goals that guide theoretical inquiry. As I mentioned at the beginning of this section, representational ideals are at the core of the practice of idealization and a systematic account of them can ultimately lead us to a more unified understanding of idealization. To that end, we now turn back to the three kinds of idealization and consider which representational ideals are associated with them.

6.3 ■ IDEALIZATION AND REPRESENTATIONAL IDEALS

Recall that Galilean idealization is the practice of introducing distortions into models in order to simplify them and make them computationally tractable. It is justified pragmatically, introduced to make a model more computationally tractable, but with the ultimate intention of de-idealizing, removing any distortion, and adding detail back to the model. Models generated by Galilean idealization are thus approximate, but carry with them the intention of further revision, ultimately reaching for a more precise, accurate, and complete model. The ultimate goal of Galilean idealization is complete representation; its representational ideal is thus COMPLETENESS.

Minimalist idealizers are not interested in generating the most truthful or accurate model. Rather, they are concerned with finding minimal models, discovering the core factors responsible for the target phenomenon. Minimalist idealizers thus adopt the representational ideal 1-CAUSAL, the ideal that says the best model is the one that includes the primary causal factors that account for the phenomenon of interest, up to a suitable level of fidelity chosen by the theorist. While minimalist idealizers may sometimes look like they are adopting SIMPLICITY, this is almost always inaccurate, because theorists engage in minimalist idealization to really understand how the target phenomena work and why they behave the way that they do. This requires finding the causal factors that really do make a difference, not a model that simply can reproduce the phenomenon qualitatively.

Like Galilean idealization's representational ideal, minimalist idealization's ideal also demands the construction of a single model for a particular target or class of target phenomena. One typically engages in minimalist idealization in order to generate explanatory models. Such models tend to be ones that simultaneously unify many target phenomena into a class and identify the causal factors which really make a difference. For the class of phenomena of interest, this will mean finding a single model, despite the fact that it will leave out quite a lot of detail which accounts for the uniqueness of each target.

Finally, we can consider MMI. The biggest difference between MMI and the other kinds of idealization is that there is no single representational ideal which is characteristic of it. Pretty much any representational ideal—including 1-CAUSAL and in rare cases COMPLETENESS—can play a role in this form of idealization. MMI arises because of the existence of tradeoffs between theoretical desiderata. This suggests that not all desiderata are simultaneously maximizable, at least in a single model. Thus the most significant aspect of MMI is that it instructs theorists to construct a series of models which pursue different desiderata and are guided by multiple representational ideals.

Consider, for example, the ecological research program that is concerned with understanding predation. A cursory look at the ecological literature on predation reveals little in the way of the search for a single, best model of predation. Instead, one finds a series of models, some of which are more precise and accurate, some of which are more qualitative, some of which are suited for populations that are homogeneously distributed in space, and some of which are flexible enough to deal with complex spatial structure. This situation is the norm in theoretical ecology. As John Maynard Smith explained, "For the discovery of general ideas in ecology ... different kinds of mathematical description, which may be called models, are called for" (Maynard Smith, 1974, 1).

For modern ecologists pursuing MMI, a full understanding of the ecological world is going to depend on multiple, overlapping, possibly incompatible models. How might we justify this kind of pluralism? One possible approach is antirealist. We could argue that maximizing empirical adequacy in some cases requires the use of multiple models. Since antirealism requires only that models be empirically adequate, the use of different kinds of idealized models is unproblematic. Another response follows Levins and Wimsatt, who understand MMI as a process of piecewise approximations to reality (Wimsatt, 2007). As Levins puts it, "[O]ur truth is at the intersection of independent lies" (see Chapter 9 for more discussion about this idea).

6.4 ■ IDEALIZATION AND TARGET-DIRECTED MODELING

In the previous sections, I have discussed various ways that models can depart from reality and some of scientists' motivations for constructing these models. Let us now return to target-directed modeling and consider how idealization plays out in this context.

Recall that target-directed modeling is the simplest kind of modeling, in which we have a single model being used to learn about a single target. In Chapter 5, I was vague about the standards used to evaluate the model–target fit, saying only that different theorists can have different standards. Now we are in a position to make this claim somewhat more specific. Although there may be times when a model can be made completely veridical with respect to its target, complicated targets often require models that introduce at least some simplifications or outright distortions.

There are thus two types of idealization that are potentially relevant to target-directed modeling: Galilean idealization and minimalist idealization. When theorists intend to keep adding detail to their models until they perfectly represent their targets, despite whatever limitations the models have a present, then they are engaging in Galilean idealization. If, on the other hand, they are trying to produce a model that captures the core causal features of the target, and they have no intention of continuing to make the model more sophisticated, then they are engaging in minimalist idealization.

As I have already discussed, one cannot simply look at a model and determine if that model is a product of Galilean or minimalist idealization. One needs to know the representational ideals used to generate such a model. Moreover, if one looks at a synchronic snapshot of a theoretical episode, it might also be impossible to determine which kind of idealization is happening. The fact that Volterra used a very simple model of predation, for example, is consistent with two possibilities. He might have been trying to isolate the core population interactions, where more complicated models would just be a distraction (minimalist idealization). Alternatively, and this is more consistent with his own writings about the subject, Volterra might have seen his model as merely the first step in a series of complications that would continue to make the model more realistic (Galilean idealization).

Multiple-models idealization cannot be associated with target-directed modeling, because it requires moving beyond the simple one-model-to-one-target practice. Multiple-models idealization may retain a single, complex target, but construct multiple models for this target. This is motivated by the extreme complexity of the target as well as the tradeoffs associated with modeling. The former happens in cases where different idealizing assumptions are made with different models, and the latter because different models are generated by valuing different desiderata that trade off against one another (see also Wimsatt, 2007, for an extended discussion of getting a handle on complexity by using multiple models).

We have learned three things in this chapter: First, and perhaps most obviously, models often fall short of veridicality with respect to their targets. To put this point in more model-oriented language, models are often not highly similar to their targets. Second, idealization is intentional. Idealized models lack similarity to their targets because theorists have intentionally introduced distortions. Finally, the extent to which idealizations will remain depends on theorists' goals or representational ideals. When theorists intend to make their models as similar to their targets as possible, idealization will abate with the progress of science. However, when their goal is to capture the core causal features of their targets, to enhance generality or simplicity at all costs, or to maximize predictive accuracy, idealization may be permanent.

7 Modeling Without a Specific Target

Target-directed modeling, the practice of constructing a single model to study a specific target, is not representative of the entire practice of modeling. Indeed, theoretical research typically involves the exploration of classes of models aimed at understanding classes of phenomena, not the study of individual observations or individual phenomena. Such investigations can take several forms. One type of investigation involves the construction of models in order to study general phenomena such as parasitism or sexual reproduction. A second is when theorists construct models to study nonexisting phenomena. A third type of investigation involves studying a model with no target at all, stopping at the analysis stage of modeling. This chapter is about these three kinds of modeling, which I collectively refer to as *modeling without a specific target*. I will call these types of modeling *generalized modeling, hypothetical modeling*, and *targetless modeling* respectively. I will often speak of 'generalized models,' 'hypothetical models,' and 'targetless models,' but these phrases should be regarded as a shorthand for cumbersome expressions such as "the models generated by the practice of generalized modeling."

My discussion in this chapter will focus on how my account of target-directed modeling must be modified to accommodate these cases. The modifications have to do with the nature of the modeler's targets, and how the models relate to other real-world phenomena that are not part of their targets. For each of the three types of modeling I will ask:

1. What are the main targets of these models?
2. How can the models tell us about such targets?
3. What can the models tell us about the world, beyond their direct targets?

7.1 ■ GENERALIZED MODELING

The first kind of nontarget-directed modeling occurs when a generalized phenomenon is chosen as a target, not a specific instance of that phenomenon. This is the case, for example, when one is interested in studying the properties of evolution in general or chemical bonding in general. I call this type of modeling *modeling a generalized target*.

Modeling a generalized target is the aim of much of theoretical research in sciences dealing with complex phenomena. Consider the following example taken from a handbook intended to teach advanced students how to construct and analyze ecological models:

> ... sexual reproduction befuddles biologists to this day. Reproduction is certainly possible asexually, that is without mating, by mitosis of single cells, and by fission or fragmentation of multicellular organisms. So why is there sexual reproduction when asexual reproduction might be, in some sense, easier? Well, this kind of question invites models that offer ideas for why sexual reproduction may be somehow better than asexual reproduction. (Roughgarden, 1997, x)

Roughgarden concludes this discussion by saying that our only option here is to construct and analyze a model.

What would such a model look like? As Roughgarden says, there is no consensus on this issue. Indeed, one recent review counted 20 distinct ideas about the evolution of sex (Hurst & Peck, 1996). However, one popular idea, due originally to J. B. S. Haldane (1932), is that the genetic recombination associated with sexual reproduction allows for a greater exploration of phenotype space. This would generally be favored by natural selection because increased phenotypic diversity better equips populations to deal with changing environments. If the environment changes, and the direction of selection changes with it, a more phenotypically diverse population has more potential resources to deal with these changes.

J. F. Crow (1992) is responsible for an influential attempt to formalize this idea into a model. Crow's model begins by assuming that sexual reproduction already exists. It then focuses on a hypothetical competition between a sexual population and an asexual population living in the same environment. The populations are given an initial distribution of genotypes and are subjected to intense selection. Using this model, Crow showed that intense selection reduces the variance of genotypes (as measured by change in variance of fitness) in the asexually reproducing populations. In sexually reproducing populations, the variance is more or less conserved. This corresponds to more phenotypic diversity in the sexually reproducing population.

What happens when the environment changes? Crow writes,

> Suppose that after a few generations of selection, the entire direction is reversed: individuals with a lower fitness potential in the former environment now have a higher fitness. The variance of the asexual population continues to decrease. In fact, immediately after reversal, the reduction is greater than before because of the highly skewed shape of the distribution in the truncated tail. In the sexual population the opposite occurs. The variance is not reduced further. In fact, the immediate effect is to increase the variance because the population is now being selected in the direction of gametic equilibrium. (Crow, 1992, 172)

This means that the sexual population was better equipped to handle this (admittedly extreme) change to the environment. For this model at least, Haldane's hypothesis is correct.

We have now seen an example of a model for a generalized target system. So let's consider our three key questions: What are the model's targets? How does it tell us about these targets? And how does what it tells us about these targets relate to other real-world phenomena?

First, consider the target of a generalized model. Such a target is, by its very nature, more abstract than any specific target. A generalized model of sexual reproduction isn't supposed to be about kangaroo sex or fungi sex, but about sex itself. This is why Crow's model of the origin of sexual reproduction has no detail about any specific species in it. But nothing in the world looks like "sex in general." There is kangaroo sex, Tasmanian devil sex, and human sex, but not sex in general. Sex in general is an abstraction over these more specific kinds of sex. So how shall we understand the nature of the target of a generalized model?

A simple account of the targets of generalized models says that these targets are the shared features of a number of specific targets. For example, the target of a model of sexual reproduction is simply the set of properties shared by all sexually reproducing species. Such a target is generated by finding the intersection of the total states for each target. In other words, the theorist takes the features common to all instances of sexual reproduction to be the target of her model. Call this the *intersection* view.

When a nontrivial set of relevant features are actually shared by all the targets of interest, the intersection view is correct. In some cases, however, attributes of the generalized phenomenon are not actually shared by more specific targets. Such targets may have some, but not all, of the relevant features. Another possibility is that the features may be only similar among the specific targets, not strictly shared among them, such as the features associated with learning across many different species. In this case, theorists must abstract further, finding a higher-order property that is instantiated among the specific targets.

In either case, there are two possibilities about how a model can be related to a target depending on the level of abstraction of the model. The simpler case occurs when the model is of the same level of abstraction as its target. For example, in describing his intended target, Crow writes:

> The model discussed here would apply strictly to a sexual population in which a large number of asexual mutants appear and are otherwise the same as the original population. It should also be appropriate to a population in which sexual and asexual strains have existed for some time in a stable environment. (Crow, 1992, 172)

This makes it clear that Crow regarded his model's targets as having the same level of abstraction as the model. He intended the dynamics described by his model to actually be shared by these systems. So this simple case works just like target-directed modeling.

More generally, the simple case will apply to generalized modeling when the following criteria are met:

1. The relevant set of specific targets actually share the relevant features, such that an intersection of their sets of features is an informative generalized target.
2. A model can be constructed at the appropriate level of abstraction so that just those features can be modeled.

These criteria are met for Crow's model, since his model was a population genetic model. Such models needn't have any details about particular species' habits, or even the molecular mechanisms involved in mitosis. The same is true for much of traditional mathematical modeling.

However, many computational and concrete generalized models fail the second criterion. Such models cannot be constructed at a level of abstraction where only shared features will be modeled. They are almost always more concrete, containing feature types that go beyond what is shared between the specific targets.

For example, imagine a computational analogue of Crow's sexual reproduction model in the form of an individual-based model. Such a model would not represent the distribution of genotypic fitness in a population, but would have specific genotypes tied to specific organisms that engaged in reproduction. This would require the theorists to make assumptions about the life-cycle, spatial distribution, mating interactions, and fitness of each organism in the model. The model would no longer contain just those types of properties shared with the abstract target; it would be closer in concreteness to specific-organism targets. The same situation would arise, but to an even greater extreme, if the theorist constructed a concrete model out of model organisms.

In cases where the model has more concrete features than the abstracted target, the theorist cannot simply connect her model to abstracted targets. Instead, she must engage in what I will call *construal setting*. This means that the theorist places special constraints on her construal of the model, especially the model's assignment and intended scope. If some features of the model cannot be part of the abstracted target, the theorist must regard these features as remaining outside the model's scope. This is similar to other scope restrictions, such as choosing to ignore fractional or infinite values.

An individual-based analogue to Crow's model would need to represent a fluctuating environment by representing some particular resource of the environment as fluctuating. To compare this model to the abstract target of sexual reproduction in general, its scope would need to be restricted via construal setting. The theorist would treat the model as if it simply said that some resource was fluctuating, even though it seems to give more concrete detail. Scope restriction via construal setting thus allows the interpreted model to be treated at the same level of abstraction as the target, yet maintain structural features that are required for the model to generate its behavior such as particular computational routines or physical properties.

So far we have seen that generalized models have abstracted targets and that they are connected to these targets in a manner akin to target-directed modeling. The main difference between target-directed modeling and generalized modeling is in the level of abstraction of the target, not the model–target relationship itself. However, generalized models are usually studied to learn general information about the world. Such models have two related functions: They can generate *how possibly* explanations and they can play a role in minimalist idealization.

7.1.1 How-Possibly Explanations

Generalized models can be used to answer how-possibly questions, such as, what is a possible reason for sexual reproduction when asexual reproduction is less costly? How can discrete alleles generate what looks like continuous variation? How can the electronic properties of a carbonyl group on one side of a molecule affect the electronic environment of a hydrogen atom on the other side? Much effort in theoretical science goes to trying to answer such questions, either as part of a larger explanatory project or to guide experimentation. A good answer doesn't depend on explaining how any specific system actually works, but rather on how they might work.

For example, consider Thomas Schelling's model of racial segregation. His main result is that small preferences for like neighbors can lead to segregation. This result is robust across many investigated changes to the model including changes to utility functions, rules for updating, neighborhood sizes, and spatial configurations. It can even be generated when agents prefer to be in the minority.

One way to use Schelling's model is to try to account for the patterns of segregation in a real city. For example, in Figure 7.1, I show the population density of African Americans in Philadelphia's census tracts next to an example output of a Schelling model. Both have a 75% exposure index, meaning that 75% of neighbors are of the same race. This use of Schelling's model is an example of target-directed modeling.

However, Schelling's own use of the model was different. He tells us that he was interested in examining "some of the *individual* incentives and individual perceptions of difference that can lead *collectively* to segregation" (Schelling, 1978, 138). In other words, how is it possible for segregation to happen in a city

Figure 7.1 Left: Patterns of racial segregation in Philadelphia from the 2010 census. Darker areas correspond to higher percentages of African-Americans. Right: The output of a typical run of Schelling's segregation model using a virtual city that is 51 × 51 and two types of agents (gray and white). All agents prefer to have at least 30% same-colored neighbors.

without collective preferences for segregation? The answer is that this is possible when every individual has a small preference for similar neighbors and tries to satisfy this preference. Along the same lines, population geneticists wondered if it was possible for an allele to become fixed in a population with no selection. They showed that among other mechanisms, genetic drift alone can lead to fixation (Ewens, 1963; Kimura & Ohta, 1969). These are examples of generalized modeling to answer how-possibly questions.

Generalized modeling can also provide counterexamples to long-established how-possibly explanations. In such cases, the theorist tries to show that a how-possibly explanation we had previously accepted was mistaken, or that the mechanism we previously believed to explain a phenomenon couldn't possibly explain that phenomenon. For example, May's famous work on complexity and stability in model ecosystems showed that, contrary to received wisdom, greater complexity in ecosystems did not always lead to stability. Rather, systems with greater numbers of species can become unstable in some conditions (May, 2001). May reached these conclusions not by studying models of specific targets, but by studying extremely abstract models of population dynamics.

7.1.2 Minimal Models and First-Order Causal Structures

A second way that theorists can learn from generalized models is by investigating what Roughgarden calls *minimal models for an idea*. Such models combine a number of causal factors to investigate what kind of behavior these factors generate. A theorist starts from an idea about how some phenomenon works. She then constructs a model that captures the most important features of the mechanism, and studies the behavior of this model. The goal here isn't a how-possibly explanation per se, but the exploration of a causal mechanism using a model.

To illustrate, let's consider Craig Reynolds' *boids* model (1987) of flocking behavior. This model was designed to study how bird flocks could cohere without a master controller. The goal was to show that such behavior could arise if each bird followed a simple set of rules.

Boids-type models are individual-based models, meaning that organisms are represented separately, possess their own variables, and implement procedures themselves. The flocking individuals implement three basic rules:

> **Separation:** steer away from nearest neighbors to keep from getting too close
> **Alignment:** steer towards the average heading of neighbors
> **Cohesion:** move towards neighbors

To implement the boids mechanism in a model, theorists must create specific implementations of these rules. Such implementations explicitly define the boundaries and geometry of the virtual space, the definition of a neighborhood, and a notion of distance. Parameters are needed to specify the angles and distances for each of the three rules. Most versions of the boids model show remarkably coordinated flocking behavior.

Boids models were originally developed as a way for computer animators to make more realistic populations of birds. It has since been used in computer animation to simulate many different types of coordinated motion including the movement of fish, penguins, and bats (Reynolds, 1987). In addition, boids models have also provided a template for studying emergent phenomena in general. They are considered a paradigm case of how a few simple rules can give rise to a complex adaptive system (Miller & Page, 2007).

Minimal models for an idea can eventually become minimal models for targets (see Section 6.1.2). This happens when such models are empirically determined to be the first-order causal structures of real-world phenomena. For example, while Reynolds originally thought of his model as a way for computer animators to have better illustrations of bird movement, some theorists have proposed boids-like rules as the core causal mechanism describing the flight of real birds (Heppner & Grenander, 1990). While more complicated models of this behavior could be constructed, a boids-like model would be the outcome of an episode of modeling that took 1-Causal as its representational ideal; it would be a model that represented all of the causal factors that really made a difference to bird flocking. So another role for generalized modeling is the production of potential minimal models for targets.

In summary, generalized modeling involves the use of models to get a handle on very abstract targets, sometimes moving even beyond these targets. When model and target share the same level of abstraction, such models can be directly compared to their targets. When models are less abstract than their targets, theorists must use construal setting, interpreting their model's structure more abstractly than is suggesting by the structure's representational capacity. Generalized models can be used not only to study abstracted targets, but also to answer how-possibly questions and used as minimal models of targets (see Section 6.1.3). And they illustrate how modeling can be decoupled from specific phenomena in the world, becoming part of what scientists sometimes call 'pure theory.' The next type of modeling takes us even further away from the world.

7.2 ■ HYPOTHETICAL MODELING

R. A. Fisher was one of the fathers of the neo-Darwinian synthesis of evolutionary theory and population genetics, and he was also one of the most important biological modelers. In opening his most significant work of theoretical biology, Fisher explains part of his approach to theoretical research as follows:

> No practical biologist interested in sexual reproduction would be led to work out the detailed consequences experienced by organisms having three or more sexes; yet what else should he do if he wishes to understand why sexes are, in fact, always two? (Fisher, 1930, ix)

The sentiment is expressed a bit strongly, but Fisher's main point is that deeper understanding in biology requires learning about the properties of both actual and nonactual target systems. I will call the practice of modeling nonexistent targets *hypothetical modeling*. At the end of this chapter, I will return to Fisher's example of three-sex biology, which turns out to be more complicated than Fisher thought.

The first question about hypothetical modeling can be settled simply: What are the targets of these models? *Ex hypothesi*, their targets are nothing at all. With a little more nuance, we can say that the targets of hypothetical models are possibilities. Explaining how hypothetical models can tell us about such possibilities would require a lengthy discussion about the metaphysics of possibilities, which is beyond the scope of this book. However, more important than what hypothetical models tell us about their own nonexistent targets is what they tell us about real-world phenomena, which will be my focus in this section.

There are two kinds of things that hypothetical models can tell us about real-world phenomena, corresponding to two cases of nonexistence: contingent nonexistence and nomically necessary nonexistence. For the first, the theorist constructs a model of a target which, as a matter of contingent fact, does not exist. For the second, the existence of a target is physically impossible.[1] The rest of this section is divided into these two categories.

7.2.1 Contingent Nonexistence: xDNA

Mathematical, computational, and physical models can all be used to study the properties of contingently nonexistent systems. To illustrate, I will first consider the case of xDNA, a concrete model at the molecular scale.

In 2003, Eric Kool and his colleagues described the synthesis of size-extended analogues of the nucleosides adenine and thymine (Liu et al., 2003). Each synthesized base was 2.4 angstroms larger than naturally occurring bases because of the insertion of a benzene ring, converting the bicyclic purine ring of adenine into a three-ring structure and the monocyclic pyridimine of thymine into a bicyclic structure. The resulting molecules are called benzoadenine (xA) and benzothymine (xT). When attached to deoxyribose rings, these yield the nucleotides deoxybenzoadenosine (dxA) and deoxybenzothymidine (dxT), which are analogues of the conjugate DNA bases dexoyadenosine (dA) and deoxythymidine (dT) respectively. These are shown in Figure 7.2

After preparing the adenine and thymine analogues, Kool determined that these analogues could function as pairing partners in the fashion of DNA base-pairing; xA can be paired with T and xT can be paired with A. After they established that this pairing was possible, Kool and colleagues synthesized a ten-base pair sequence containing only xA and T. This sequence was paired with a complementary strand to form a Watson–Crick helix, which Kool called xDNA.[2]

This particular strand of xDNA, and others that were subsequently synthesized, adopts a very similar three-dimensional structure to ordinary DNA. xDNA forms a right-handed double helix, although as with naturally occurring DNA, a triple helix is also possible. While the topology is preserved, the helix is wider, about 12.9 Å instead of 10.7 Å in DNA. This wider strand accommodates additional base pairs in each turn of the helix. Like DNA, xDNA is thermodynamically stable and will spontaneously form in solution. In fact, xDNA is more stable than DNA by 5.8 kcal/mol (free energy of formation at 37°C) due to enhanced base stacking along the long axis of the strand. Finally, because of the greater π-conjugation of the extended bases, dxA and dxT are brightly fluorescent.

---

1. I present these as dichotomous alternatives, but there is a continuum of contingency. See Mitchell, 2000, for a discussion of systems that range from the physically impossible to the contingencies generated by true accidents of history.

2. Each base pair consisted of one extended and one standard base, so that each segment of the helix was the same width. In a subsequent paper, Kool reports the synthesis of a complete eight-base xDNA helix (Gao, Liu, & Kool, 2005).

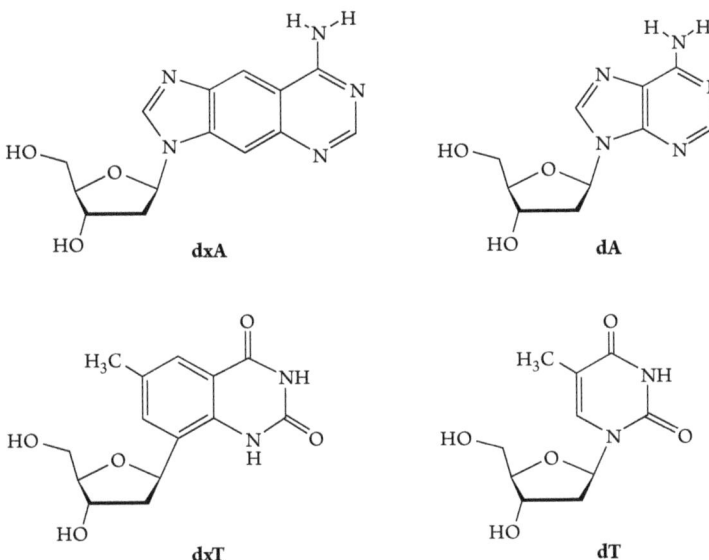

Figure 7.2 Structures of the size-expanded nucleosides benzoadenine and benzothymine. From Haibo Liu, Jianmin Gao, Stephen R. Lynch, Y. David Saito, Lystranne Maynard, and Eric T. Kool, "A four-base paired genetic helix with expanded size," *Science*, 302/5646 (31 October 2003): 868–871. Reprinted with permission from AAAS.

xDNA has considerable practical uses. For example, it can be used as a probe of naturally occurring DNA sequences due to its fluorescence. But Kool saves the most interesting conclusion for his final discussion. He writes:

> We conclude that a genetic base-paired molecular framework is not limited to the size of the DNA helix that evolved in terrestrial biology. It is noteworthy that one significant difference of this expanded-size genetic system, relative to the natural one, is its increase potential for encoding information. Pairing of our expanded bases with four natural bases in all complementary combinations would be expected to yield eight base pairs of information encoding ability (Liu et al., 2003, 871).

On the basis of this physical model, Kool concludes that the familiar base-pairing system of DNA is only one possible genetic system; DNA's ubiquity is contingent.

To further draw this point out, we can ask somewhat anthropomorphically whether nature "chose" DNA as the basis of heredity, or if DNA is a historical accident. The xDNA model certainly can't settle this question, but it does suggest that a well-functioning genetic system can be constructed based on xDNA.

So, in all likelihood, the evolutionary explanation for the use of DNA instead of xDNA in our genetic system is a frozen accident.[3]

Contrast this with what most chemists believe about nature's choice of phosphates for DNA's backbone. In an influential article, Frank Westheimer argued that phosphates are uniquely suited to their biochemical role in DNA, RNA, ATP, intermediate metabolites, coenzymes, and so forth. Focusing on DNA, we might ask whether, like the base-pair system, nature's use of phosphates in the DNA backbone is a contingent matter. Westheimer argued that there is no known viable alternative to phosphates because they can simultaneously form ester bonds, permitting DNA strands to be stable yet easily cleaved. At the same time they can be charged, helping the backbone to survive indefinite periods in a water solution. This combination of lability and charge is found in no other molecule. Thus, Westheimer argues, nature chose phosphates by some chemical analogue of natural selection (Westheimer, 1987; for further discussion of Westheimer's arguments and recent empirical challenges, see Parke, in press).

The contrast between the phosphate backbone and DNA nucleosides is illuminated by xDNA. Kool's molecular model shows that the nonexistence of xDNA is a contingent fact. Terrestrial biochemistry could have been based on xDNA, while it is unlikely it could have used a DNA-like system without a phosphate backbone. This is an important lesson that can be taught by modeling a nonexistent phenomenon: This target phenomenon might well have existed.

Physical models like xDNA can be high-fidelity models only of contingently nonexistent systems. If one wants to study the properties of targets that are nonexistent for nomological reasons, then one must either adopt very low standards of fidelity, or else move into the realm of mathematical modeling. The next two examples illustrate how theorists can model necessarily nonexistent targets, and what can be learned about the world from such models.

7.2.2 Impossible Targets: Infinite Population Growth and Perpetual Motion

Computational and mathematical modeling can let us study systems which cannot physically be manifested in the real world. As a very simple example, consider the populations described by exponential growth models. One such

---

3. Of course, this model alone cannot be used to definitively draw these conclusions. We would need to see how efficiently xDNA can replicate, whether the appropriate transcription system could be set up, and so forth. These are nontrivial things to show, but roughly the same kind of chemical biology could be used to show them.

model is described by the following instantiated model description:

$$\frac{dN}{dt} = .01r \tag{7.1}$$

Whatever the starting value for N, the population size or population density, this population will increase exponentially. No real population can sustain such growth in the actual world because ultimately space and resources are limited. Yet in his classic work on theoretical ecology, John Maynard Smith (1974) devotes a whole chapter to this family of models. What can we possibly learn from a model such as this one where its target system cannot possibly exist in our world?

While no real population can have long-term exponential population growth, the model may be a reasonable approximation of population growth in the short term. More importantly, there are several things that can be learned about real-world targets by studying this model, at least when considered with a wide scope and moderately high fidelity. First, one interesting reason to study this model is because it can be a submodel of a more complex model which describes population dynamics with higher fidelity. For example, exponential population growth is the first term of the prey equation used to describe the Lotka–Volterra predator–prey model. Understanding this model on its own may help us better understand the contribution it makes to the Lotka–Volterra model and other models in which it is embedded.

Another reason to study this model is that it describes the limiting behavior of a more realistic family of models: the logistic growth models. These models are described by equations of the general form:

$$\frac{dN}{dt} = rN\left(1 - \frac{N}{K}\right) \tag{7.2}$$

In models described by this equation, the environment has a maximum number of organisms it can support—the *carrying capacity*, designated by K. As K $\to \infty$, this equation becomes the exponential growth equation. So while the exponential growth model's target is impossible, it can demonstrate the limiting behavior as carrying capacities get larger and larger. In this particular case, logistic models are easy enough to understand without a detailed analysis of exponential models as a limiting case, but this isn't always so. In many cases, understanding models with impossible targets is a necessary prelude to understanding more complicated models that include them as special cases.

Models with no possible real-world target can also play a role in scientific explanations. Many philosophers have regarded generality as an important desideratum of scientific explanation (e.g., Hempel, 1965; Friedman, 1974; Kitcher, 1981; Strevens, 2004). Thus if one wanted to explain why any population with property P will also have property Q, then according to these theories, an especially powerful explanation of this fact would transcend the actual and also apply to nonactual populations. Thus theorists study models of impossible targets to learn about and explain actual systems, not just the impossible targets themselves.

My final example of modeling nonexistent targets takes us even further away from the realm of possibilities, and thus cannot be a template for models of real-world targets or approximately explain the behavior of these targets. The best examples of such models are ones that violate the laws of nature.

A perpetual-motion machine is a machine that can produce work indefinitely. Most thermodynamics textbooks explain that the construction of an actual perpetual-motion machine is impossible because such a machine would violate the second law of thermodynamics. However, one can construct a model of a potential perpetual-motion machine and use it to deepen our understanding of why such machines are impossible. Such a model could also tell us which laws of nature would have to change in order for such a machine to be constructible. The most vivid illustration of this is called the *ratchet and pawl machine*, or sometimes *Feynman's Ratchet*, and it is depicted in Figure 7.3 (see Feynman, Leighton, & Sands, 1989, for Feynman's original description and depiction of this model).

The ratchet and pawl machine is a model whose components are at the same temperature, but which nevertheless can extract work from its heat reservoir. The model machine can be described as follows. Two boxes are connected by an axle. Vanes are attached to one end of the axle, which is enclosed in box 1. Box 1 has temperature T_1. A ratchet is attached to the other end of the axle. The ratchet is placed in box 2 with temperature T_2. The ratchet wheel has teeth

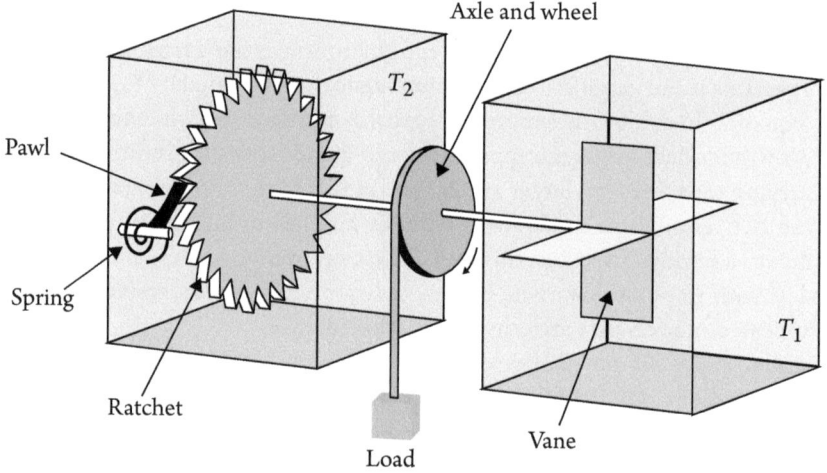

Figure 7.3 Feynman's ratchet illustrated by Zhanchun Tu. Courtesy of Professor Tu.

angled in one direction which catch on a protrusion mounted to the side of the box (the pawl). This ratchet allows the axle to turn only in one direction. Finally, between the two boxes in the middle of the axle is a wheel. Wrapped around the wheel is a cord tied to a weight. Whenever the wheel turns and the weight is lifted, the machine is doing work.

There is one final crucial part of the ratchet and pawl machine. The pawl must have a damping mechanism in order to be effective. If its movement was perfectly elastic, then any time it was raised for the ratchet wheel to turn in the right direction, it would continually bounce on the ratchet wheel, never dissipating its energy. If this happened, the wheel would be able to turn more or less freely, and the ratchet would not be effective. So we will assume that the pawl is damped and releases heat to the box 2 system.

Our ratchet and pawl machine seems to be an excellent model of a perpetual-motion machine. Here is how it could work: If box 1 is filled with a gas at temperature T_1, the vanes will be struck by the gas molecules. Normally, this would have no net effect because the molecules are equally likely to strike the vanes on one side or the other. However, because the vanes are attached to the axle which is attached to a ratchet, the only effective collisions will be ones in a particular direction, the direction "allowed" by the ratchet.

Since the axle will turn only in one direction, the wheel attached to the axle will only turn in one direction. Moreover, since the movement of the axle in one direction is spontaneous, due only to the action of molecular collisions, we can just sit back and let the machine operate indefinitely. It looks like we have a machine which will continually extract work—in this case raising a weight attached by cord to the central wheel—from the gas in box 1. So we seem to have a model of a perpetual-motion machine, or at least a model description of one.

But there are no perpetual-motion machines! So hopefully what we can learn from this model is why the machine wouldn't work in our world. We can do so by applying some very basic physical principles from our world to the model.

In order to apply these principles, consider in the abstract how the perpetual-motion machine is working. It takes energy from box 1 and transfers it to the wheel where this energy is used to do the work of lifting the weight. The energy it is removing from box 1 must be thermal energy, for the machine works by turning molecular collisions into work via the vanes, axle, and wheel. The way we keep track of thermal energy in a system is by measuring its temperature. So we should ask ourselves: what temperature is box 1? Or more importantly, what are the relative temperatures of box 1 and box 2?

There are three possibilities: Either box 1 and box 2 are the same temperature, box 1 is higher in temperature than box 2, or box 1 is lower in temperature than box 2. First, consider the situation where both boxes are at the same temperature. Heat from box 1 will be converted to work in the standard fashion. In box 2, every time the ratchet wheel turns, the damping mechanism absorbs the energy from the collision of the pawl and the wheel teeth. The absorbed energy is converted to heat and the temperature of box 2 begins to rise. As the temperature rises in the box, there are more molecular collisions with the ratchet wheel and, importantly, the pawl. Energetic collisions with the pawl from the correct angle will lift the pawl and let the ratchet wheel turn backwards, reversing the work performed by the machine. So in the case where $T_1 = T_2$, the ratchet and pawl machine cannot effectively function as a perpetual-motion machine.

For parallel reasons, the case where $T_1 < T_2$ is a nonstarter. This is precisely the situation that ultimately defeats the case with equal temperature. Finally, the case where $T_1 > T_2$ is a standard thermodynamic machine that does work by pumping heat from a heat sink to a cold sink. Unfortunately, this machine will not work perpetually. As the machine works, box 1 and box 2 will eventually come to have the same temperature, since heat is being transferred from box 1 to box 2. Once box 1 and box 2 are equal in temperature, then this becomes an example of the $T_1 = T_2$ case, and will ultimately stop functioning or even run backwards.

The machine could continue to operate in the first and third cases, of course, if heat could be removed from box 2. But then some additional, external process would be required and the machine wouldn't be a perpetual-motion machine. What would need to happen for this model to really work in a world like ours is for the physical laws to be different. In particular, energy conservation would have to be violated and some of the transferred energy would have to simply disappear rather than being converted to heat.

By exploring physically impossible cases with models such as the ratchet and pawl machine, we gain deeper appreciation of the consequences of our laws of nature. In this case, for example, we learn about the nomological dependence between energy conservation and perpetual motion machines.

Having now looked at several possible relationships between models of nonexisting targets and real-world systems, we can reexamine and reassess Fisher's claim that theorists ought to study models of nonexistent targets. Why should theorists who are primarily interested in studying what is actual try to understand what isn't actual? The answer to this question cuts deep into the heart of theoretical practice: Theorists ultimately aim to partition the space of possibilities. They aim to understand what is possible, what is impossible, and why.

Moreover, complete understanding of actual systems may require having substantial counterfactual knowledge, what I have called p-generality in Chapter 6. Such knowledge goes beyond the actual, telling us what the world would be like if the model's structure and behavior were instantiated in our own world. For contingent hypothetical models, we learn that a different history could have led to a different reality. From models of nomologically impossible systems, we can learn why our world cannot have the model system and what laws of nature would have to change in order to make this merely a contingent fact. Gaining this knowledge can be part of a general project of learning about possibilities, or can be applied as the counterfactual background required to explain specific phenomena in the world.

7.3 ■ TARGETLESS MODELING

In the final type of modeling that I will discuss, no target is chosen at all. The only object of study is the model itself, without regard to what it tells us about any specific real-world system. This type of modeling is most akin to pure mathematical analysis.

One set of models that is frequently studied for their own sake, and perhaps for the sake of what they tell us more broadly about computation, are *cellular automata* (Ilachinski, 2001). Such models consist of an array of *cells*, which can each be in one of some number of states. Transition rules determine how the states change, and these rules typically depend on the states of neighboring cells.

A simple and well-known version of a cellular automaton is the Game of Life, first studied by J. H. Conway in 1970 (see Gardner, 1970). This game consists of an infinite two-dimensional array of cells that can be in one of two states: alive (1) or dead (0). Neighborhoods are defined using the Moore neighborhood definition, the eight cells adjacent to a center cell. Each time-step of the game involves evaluating the transition state for each cell according to the following rules:

1. If a live cell has fewer than two live neighbors it dies.
2. If a live cell has two or three live neighbors it doesn't change state.
3. If a live cell has more than three neighbors it dies.
4. If a dead cell has exactly three neighbors it transitions from dead to alive.

After an initial distribution of live and dead cells is specified, the computer evaluates the rules at each time-step, updates the cells, and evaluates again. For obvious reasons, the only practical way to examine the game's behavior is using a computer, and this requires using finite analogues of the game, although as described the game is played on an infinite array.

One of the most dramatic visual results of the game is the existence of patterns that have life-like behavior such as movement, reproduction, "eating" other patterns, and so forth. For example, the pattern called a *glider* depicted in Figure 7.4 moves across the board diagonally from top left to bottom right.

Much of the study of Conway's Game of Life has been for recreation; this game and related cellular automata had a large cult following among

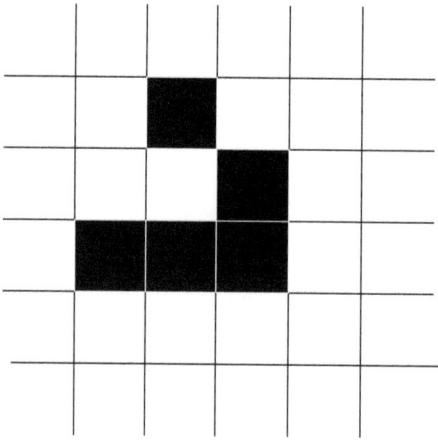

Figure 7.4 A glider in the game of life. This structure persists and will "move" across the game board.

computer enthusiasts in the 1970s and 1980s. However, more recently, computer scientists have studied the game more formally. First, it was proved that, with an infinite grid, the game was Turing complete—it can be used to compute any computable function (Berlekamp, Conway, & Guy, 1982; Gardner, 1983). This lead researchers to investigate how a Turing machine could actually be created within the game. A finite Turing machine was implemented in the game by Rendell (Rendell, 2002; see also Poundstone & Wainwright, 1985), and this machine has been used to perform nontrivial computations.

Since targetless modeling involves constructing and analyzing models for which there is no target, studying them is in many respects a kind of abstract direct representation (Weisberg, 2007b). Although typical cases of abstract direct representation involve the study of empirically discovered phenomena, in targetless modeling, the object of study is a constructed model. Nevertheless, the development of such models has often been motivated by ideas about the way that the world might work, be it crystal growth, evolution, or the emergence of consciousness, even if they are not intended to be models of such phenomena. Thus there are at least two ways that cellular automata can come in contact with real-world phenomena.

The first way we might call metaphorical. Various authors have argued that the Game of Life and related cellular automata can give us insights about biology (Eigen & Winkler, 1983; Kauffman, 1993; Langton, 1995; Ermentrout & Edelstein-Keshet, 1993; M. A. Nowak, 2006), physics (Wolfram, 1983; Vichniac, 1984; Rothman & Zaleski, 2004), and other sciences, in part by helping us to sensitize our imagination so that we learn how to notice things we might have missed otherwise.

For example, Dennett discusses the interesting fact that when we begin thinking about the Game of Life, we start by describing a grid, cells, and the rules for each cell. But fairly soon we are talking about gliders, glider guns, and all sorts of patterns at a higher level than individual cells.

> Note that there has been a distinct ontological shift as we move between levels; whereas at the physical level there is no motion, and the only individuals, cells, are defined by their fixed spatial location, at this design level we have the motion of persisting objects; it is one and the same glider that has moved southeast [...]. Here is a warming-up exercise for what is to follow: should we say that there is real motion in the Life world, or only apparent motion? The flashing pixels on the computer screen are a paradigm case, after all, of what a psychologist would call apparent motion. Are there really gliders that move, or are there just patterns of cell state that move? And if we opt for the latter, should we say at least that these moving patterns are real? (Dennett, 1991, 39)

Dennett thus uses the relationship between the apparent ontology of the Game of Life world—gliders, puffers, beacons, toads, and so on—and the very simple underlying states and transitions of the game. It turns out to be very difficult to say whether the real patterns exist. Dennett uses this to draw larger conclusions about the metaphysics of mind. But the model itself isn't used in this discussion, merely the insights drawn from investigating the model.

A second way we can learn about the world from targetless models is more direct. Sometimes, targetless models can inspire a more general modeling framework that can be used for target-directed modeling. For example, my colleague Ian Lustick (2011) has used a very complicated Game of Life-like cellular automaton to study stability and unrest in countries. Rather than using a square grid, Lustick uses dimensions for the grid that approximately correspond to the shape of a country. Cells of the grid represents pockets of political identities, and transition rules are based on constructivist identity theory, the idea that individuals' social and political identities are constructed from the ones they see around them, especially among people with considerable social influence.

In summary, models without targets are primarily studied for their intrinsic interest. However, when these models generate a productive literature, it is often only a matter of time before researchers begin to find new uses for them. Such was the case of cellular automata used to study political unrest.

7.4 ■ A MOVING TARGET: THE CASE OF THREE-SEX BIOLOGY

I have now discussed the three main types of modeling without a specific target. Before closing this chapter, I want to return to Fisher's famous claim about the evolution of sex. He suggested that three-sex biology was an example of a model with a nonexisting target, but it turns out that the story is considerably more complicated, and it nicely illustrates the fluidity of modeling practice.

Fisher was convinced that three-sex biology was a nonexistent system and so modeling it would be only of theoretical, not practical interest. The first complication to this simple picture is an empirical one: Some species seem to have more than two sexes.

In mammals, a simple way of distinguishing the sexes is by the size and quantity of sex cells. Males produce many small sex cells, females produce few large sex cells. In *isogamous* species, such as many kinds of ciliates and fungi, all sex cells are the same size. In these species, sex must be determined directly by who can mate with whom. This leads to the idea of a *mating group*, a group of organisms of the same species that cannot mate among themselves. When the number of sexes is counted as the number of mating groups, it turns out that some species have literally thousands of sexes (Raper, 1966; Casselton, 2002). So without looking into the matter further, Fisher's claim that there are no three-sex systems is empirically suspect.

Theoretical work on three-sex biology also puts pressure on Fisher's claim. As theorists began writing down explicit genetic models of different kinds of sexual systems, it became clear that Fisher's reference to "three-sex biology" was ambiguous. The first ambiguity concerns mating combinatorics. If there are three sexes, A, B, and C, can any pair mate? Are only some combinations allowed such as A/B, B/C? Or does mating somehow require a three-way fusion of sex cells? It is very likely that Fisher was thinking of a three-way fusion and argued that this was very unlikely or impossible, just as three-way collions of gas molecules are very improbable. But this is just one way of having a three-sex system.

Most subsequent authors have used the simplest version of mating combinatorics, where any two non-same mating types could mate (for further discussion, see Bull & Pease, 1989). As theorists began investigating different mating combinatorics, along with different assumptions about the life histories of these organisms, it began to become clear how far off the mark Fisher's original claim had been. In one of his papers on the subject, Laurence Hurst sketches a model of the evolution of mating types.

> Now consider the situation in which there are two mating-types. A gamete with a novel third type will be able to mate with all other gametes when it initially appears in the population. Assuming there is some cost to the finding of mates, one would hence expect that such a mutant should easily invade. The same is true for the invasion of any novel mutant into a population with more than two mating-types. (Hurst, 1996, 415)

What conclusions should we draw from this? Hurst writes:

> It is therefore expected that either there should be zero mating-types, or, if for whatever reason mating types have evolved, that the number of mating-types should tend towards infinity [...]. Two sexes is the least expected condition. It is therefore remarkable that so many organisms have two sexes. (415)

So now the question has changed. Instead of asking what would nonexistent three-sex mating systems look like, the new questions are: Why aren't all mating systems n-sexed? Why are so many systems two-sexed? These are open research questions to this day.

A final subtlety takes us back to my initial association of mating types with sexes. I made this association quickly and without argument, but there are clearly other ways to think about how many sexes a species has. Parker has argued that there are three reasonable ways to define a species' number of sexes. The first is by counting the number of different gametic types, which is equivalent to my associating mating types with sexes. The second possibility is to count how many gamete types must combine to form a fertile individual. If we use this definition, then the fungi with thousands of mating types would still be counted as having only two sexes. Finally, we can count the number of gametic types required to stabilize and prevent extinction of a sexually reproducing population (J. D. Parker, 2004).

As it turns out, there are some recently discovered cases of social insects that require the contribution of three distinct sexes in order to prevent the extinction of the colony. In normal haplodiploid insect systems, such as most species of ants, males and females are produced in different ways. When the queen lays an egg, if it remains unfertilized, then the egg develops into a male. However, if the egg is fertilized by a male, then the egg develops into a female. Whether this female is a worker or a queen is determined by environmental factors.

In southwestern New Mexico, biologists have discovered populations of *Pogonomyrmex barbatus*, the red harvester ant, and *Pogonomyrmex rugosus*, the rough harvester ant, that have hybridized. This hybrid system contains individuals who are clearly genetically marked as having come from one or the other species. These genetic differences are important for reproduction. Males are produced in the standard, asexual way. When the queen mates with males from the original gene pool, she produces queens, but not female workers. When she mates with males from the other gene pool, she produces workers, but not queens. For the colony to stay healthy and persist through time, the queen must mate with males from both original populations, that is, two different kinds of males. If she doesn't, the colony will die out. Thus, this is a system that requires mating among at least three sexes (J. D. Parker, 2004).

So what should we make of Fisher's famous quotation about three-sex biology? He was clearly mistaken that three-sex systems do not and cannot exist, at least under some definitions of sexes. However, the original sentiment behind his quotation was correct and is really the central point of this chapter: Theoretical science involves far more than building a single model to represent a single target. While theorists may sometimes choose to focus in this way, more often than not, they entertain many models and many targets. This highly varied practice can have many outcomes—from constructing an understanding of a single target, a class of targets, a general phenomenon, or even an impossible system. This is why I have characterized modeling from the beginning of the book as the construction and analysis of models, which can *potentially* represent targets. With this much expanded picture of modeling in mind, I will now return to the relationships between models and their targets.

8   An Account of Similarity

In this chapter, I will develop an account of the model–world relationship that is compatible with my analysis of modeling. This account is intended to be sensitive to how scientists represent the world with models and to how their representational goals and ideals shape the standards of fidelity that they apply to their models. Extant accounts of the model–world relationship tend to either be formal treatments, which draw on model-theoretic properties such as isomorphism, or informal treatments, which rely on the notion of similarity. In contrast, I will develop a somewhat formal account that is similarity-based.

As with all similarity-based accounts of model–world relations, I accept from the outset that there are many relations which can hold between models and the world and that exactly which relation is intended to hold is a matter of context and theoretical interest. In other words, model–world similarity is always a similarity in certain relevant respects and degrees. My account of the model–world relationship respects this fact and doesn't try to identify a single relation as the gold standard for representation. Rather, I will show how the model–world relationship depends on theorists' interpretations of their models, the background knowledge and practice of their research community, and their research goals.

8.1 ■ DESIDERATA FOR MODEL–WORLD RELATIONS

Throughout this book, I have explored various aspects of the practice of modeling. I have shown that models are constructed for many different reasons. Because of that, the evaluation of models is made with respect to many different standards. In this chapter, I will give an account of the model–world relationship that has the flexibility to accommodate the complexities of this practice. As a first step towards this goal, this section will present eight desiderata that an account of the model–world relationship should have. The remainder of this chapter will be concerned with developing an account that can meet all of them.

The first desideratum, called MAXIMALITY, is a formal one, having to do with the similarity relation's boundary values. A model is maximally similar to itself and to any target that shares all of its properties. All other nonidentical models will be less similar to each other or to another target. More formally, for two nonidentical models a and b and model–world relation r, $r(a,b) < r(a,a) = 1$.

Many classical discussions about the model–world relation supposed that this relation is a model-theoretic analogue to truth. It should explain how models accurately reflect the world's structure. This provides us with a starting point for an account: Good models will tell us many true things about their targets, and bad models will not. However, since models can tell us a greater or lesser number of true things depending on their degree of idealization, this relation should come in degrees and be representable on a scale. Thus, the second desideratum of a model–world relationship is that it is SCALAR.

Throughout the examples of modeling I have discussed in this book, comparisons between models and targets were made between different types of properties, including static patterns, dynamic patterns, and causal structures. For example, Lotka–Volterra models are thought to be good models when they share oscillations in species abundance with their targets. The Bay model shares, in scale, topography and tidal cycle with the Bay. It also shares certain key force ratios, such as the ratio between inertial forces and gravitational forces, with its target. So the model–world relation ought to allow for any of these kinds of features to play a role in model–target comparisons; it must allow for RICHNESS in structures that are to be compared.

Some of these patterns, properties, and structures are quantitative, but others are qualitative. For example, if one compares Schelling's model of segregation to the segregation pattern of a real city, then one will compare the fact that the model has racially segregated clusters to the fact that the city has racially segregated clusters. With such a simple model, no one would think a quantitative comparison was reasonable, yet there still may be cases where it is a good model of actual segregation patterns. Thus the model–world relation must allow for QUALITATIVE comparisons.

Another theme throughout this book has been that models are often highly idealized. Such models neither truthfully describe their targets, nor are they even intended to truthfully describe their targets. Nevertheless, idealized models can tell us many true things about their targets. An account of model–world relations should say more than "this model does not accurately represent the target" or "this model represents this target in an idealized fashion." It should be able to distinguish more successful instances of representation from less successful ones. I will call this the IDEALIZATION criterion.

My examples have also shown that model–world relations depend on theorists' explicit and implicit scientific goals. For example, how well the Lotka–Volterra model captures the properties of a particular target will depend on the modeler's fidelity criteria and representational ideals. Volterra thought he was giving a reasonably accurate model about the dynamics of Adriatic fisheries. Now we tend to think of the Lotka–Volterra as a how-possibly model or a pedagogical model. The dependence of the model–world relationship on context is the sixth desiderata for that relation, which I will call the CONTEXT desideratum.[1]

Since different theorists can have different fidelity criteria for a single model, or can apply a particular model in different contexts, then an account of the model–world relation should allow for the possibility of extraempirical disagreement about the goodness of a model. But the account should also be able to help diagnose such extraempirical disagreements, locating the sources of disagreement in context, use, and weighting of various features of the model. I call this desideratum the ADJUDICATION criterion.

Throughout the book, I have tried to describe the practice of modeling at the epistemic level, which is cognitively accessible to theorists. Similarly, I think that an analysis of the model–world relationship should reflect judgments that scientists can actually make, as opposed to asserting that the relation holds between inaccessible, hidden features of models and targets. This is clearly a modal desideratum, because in many cases theorists won't necessarily articulate the grounds for the judgments of similarity—the judgments are just made. Nevertheless, when it matters, such as in cases of disagreement, theorists should be able to work out[2] the grounds for their similarity judgment. This is the criterion of TRACTABILITY.

In the next section, I will present some of the classical accounts of model–world relations in the philosophy of science literature and argue that they fall short on a number of these desiderata.

8.2 MODEL-THEORETIC ACCOUNTS

The traditional account of the model–world relation comes from the semantic view of theories (Suppes, 1960a, 1960b; Sneed, 1971; Suppe, 1977b, 1989; van Fraassen, 1980; Lloyd, 1994). In these views, models are related to phenomena in the real world via the model-theoretic relation of *isomorphism*, although some proponents of the semantic view have weakened the requirement to *homomorphism* (Lloyd, 1994; see also Miller & Page, 2007) or *partial isomorphism* (da Costa & French, 2003). These views all take mathematical models (they say little about computational and concrete models) to be more or less equivalent

---

1. Like Stalnaker (2002), my notion of context is subjective in the sense that it is a function of scientists' shared beliefs, as opposed to a more objective Lewisian notion in which context is constituted from features like the speaker, addressee, time, and location of the utterance. In some ways, my notion of context is even more subjective than Stalnaker's. Stalnaker's account of what is said relies on the assumption of shared beliefs among the speakers. But in a theory of model–world relations, context refers only to the theoretical goals and representational ideals of a scientist or scientific community, not on mutual knowledge or beliefs.

2. This notion of 'working out' is very much like Grice's *calculability assumption* (Grice, 1975, 1981; Davis, 2010).

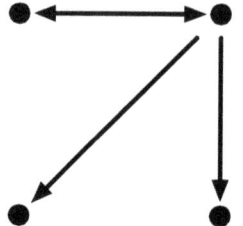 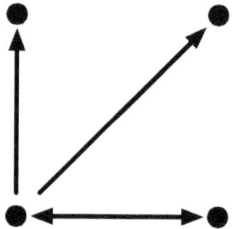

Figure 8.1 These two finite graphs are isomorphic to one another.

to logicians' models. Hence they account for the model–world relation using tools which are appropriate for logician's models.

Isomorphism is a mapping between two sets that preserves structure and relations. Formally, an isomorphism is a bijective map between two sets such that the mapping function f and its inverse f⁻ are both homomorphisms, structure-preserving maps between these two structures. Figure 8.1 shows the simple example of two directed graphs that are isomorphic to one another.

Now let's see how this account scales up to a real scientific example. Consider the use of a damped harmonic oscillator model to describe the motion of a real spring. A damped harmonic oscillator model is represented by the following equation:

$$\frac{d^2x}{dt^2} + 2\zeta\omega_0\frac{dx}{dt} + \omega_0^2 = 0 \tag{8.1}$$

where $\omega_0 = \sqrt{\frac{k}{m}}$ is the angular frequency of the oscillator, $\zeta = \frac{c}{2m\omega_0}$ is the damping ratio, and c is the damping coefficient.

To compare the predictions of this model to some target, we need to begin by specifying values for the parameters k, c, and m. This provides us with a complete set of trajectories through state space corresponding to different initial displacements of the spring. From this set, we can specify the empirical substructure of the model, that part of the model that describes measurable quantities such as the displacement of the spring through time. Formally speaking, this would constitute a set of ordered pairs {x, t} where x is the displacement of the spring and t is time. Since we are restricting ourselves to the empirical substructure, all values for time would be positive and, presumably, we would consider only some time interval of interest. According to the semantic view, we would have a good model of a spring when there is an isomorphism between this set of ordered pairs and a set of ordered pairs of empirically determined displacements and times.

Despite its ubiquity across discussions of model–world relations, the isomorphism account fails to meet many of the desiderata I outlined in the last section.

Of these failures, perhaps the most significant is its failure to meet the IDEALIZATION desideratum. To see how it fails, consider the simple, linear harmonic oscillator, which is similar to the oscillator described above, but which lacks the damping term.

$$m\frac{d^2x}{dt^2} + kx = 0 \qquad (8.2)$$

This model is used in many physical sciences to describe periodic motion, such as the swing of a clock's pendulum or the vibration of a spring. Relative to all such targets, the model is highly idealized because all springs and pendula are damped, meaning that they eventually stop oscillating because of energy loss due to friction. Because the model has no damping term, it exhibits indefinite oscillations of constituent amplitude. Similarly, the model has no provision for out-of-plane or rotational motion; all of its motion is linear and in one dimension. Yet covalent bonds, which are frequently modeled with harmonic oscillators, exhibit both of these types of motions as components of their oscillations. Harmonic oscillator models of bonds are thus highly idealized relative to their targets.

Despite their high degree of idealization, they still seem to be informative about some of their targets' properties. The reason that these models are thought to be valuable, if limited, is because they provide a simple way of characterizing the oscillatory character of targets. This fact, however, cannot be accommodated by the isomorphism account. "Oscillatory character" is not a structural property of a model; it is an interpretation of a pattern exhibited by the model. Thus there is no bijective mapping from model to target with respect to this property. And so proponents of isomorphism would have to say that these models cannot be used to describe such systems, despite their apparent utility.

Moreover, it is often impossible to find isomorphisms between the mechanistic properties of models that are supposed to explain the behavior of targets. Consider Schelling's model of segregation as applied to the segregation patterns in a real city. Insofar as the model can explain segregation in a real city, it does so by showing how a certain set of preferences and behaviors can lead to segregation. Model agents have very simple utility functions, very simple behaviors, and live on a perfectly regular grid. Real agents have complex and heterogeneous preferences, elaborate behaviors, and live on real streets and in real houses. The Schelling model's city and citizens are therefore isomorphic only to an imaginary city, but not to any real city. But an account of the model–world relationship is supposed to explain how the dynamics of Schelling utility functions and Schelling behavioral rules in a Schelling city can represent and explain, say, segregation in Philadelphia. Isomorphism cannot capture these relationships,

Figure 8.2 Left: The equilibrium state of a typical run of Schelling's segregation model. Right: A linear transformation of this state, which is isomorphic to the original.

because the structural features of Schelling's model cannot be directly mapped onto Philadelphia.

Ironically, while there seem to be too few of the right kind of isomorphisms between idealized models and their targets, the opposite is also true: there are also too many isomorphisms of the wrong kind. Since isomorphism is a mapping of structure onto structure, isomorphism is insensitive to intension; all of the interpretation of the model's structure is irrelevant to this relation. Since all that remains is structure, there will always be an infinite number of structures that any model can be mapped to. The problem is that this set of structures is unlikely to include the target system if there are idealizations in the model. All linear transformations of the equilibrium state of the Schelling models are isomorphic to the equilibrium of the Schelling model (see Figure 8.2), but none will contain some actual city that we would like to apply the model to (Hendry & Psillos, 2007; Singer, 2008).

Can proponents of model-theoretic accounts respond to these problems? Recent work on the *partial isomorphism* account (Bueno, French, & Ladyman, 2002; da Costa & French, 2003) takes some steps in the right direction. This account says that the model–world relation is tripartite. It divides a model's structure into three substructures: the part that is isomorphic to the target, the part that is not isomorphic to the target, and the part that is "left open" with respect to the target. A model is partially isomorphic to its target when a substructure of the model is isomorphic to a substructure of the target. This has several advantages over isomorphism-based accounts. For our purposes, the most important advantage is that it helps to account for cases of idealization (French & Ladyman, 1998).

For example, consider the idealization of elastic collisions that is associated with the ideal gas model. This idealization says that when two particles of the gas collide, the pair maintain their combined kinetic energy after the collision. This is not true for molecular gases (hydrogen gas, oxygen gas, water vapor, etc.) because when they collide, some kinetic energy is transferred to the molecules' internal degrees of freedom (internal rotations and oscillations). However, this idealization is not required for many ideal gas model-based explanations and, in these cases, the idealized model can be confined to the nonisomorphic substructure without any loss.

But this is not the case in general. In many of the cases I have discussed in this book, it is the idealized features themselves that are supposed to be representations of targets' features, and hence part of the explanation of the behavior of such targets. For example, the Schelling model's idealized features, such as agents' utility functions and spatial distribution, are the very things that represent properties of real people and do the model's explanatory work. Yet on the partial isomorphism account, they would have to be part of the nonisomorphic substructure because they don't match real people's properties. Hence, they could neither represent real people nor serve as an explanation of segregation in a real city (for other criticisms of how the partial isomorphism account treats idealization, see Pincock, 2005).

How about some of the other desiderata for model–world relations? Isomorphism accounts completely fail to meet the SCALAR desideratum. Models are either isomorphic to their targets, or they are not. Partial isomorphism accounts fare better, because the three-part structure of this relation would allow for the construction of a scale which says something about the relative size of the isomorphic and nonisomorphic substructures. Partial isomorphism proponents could potentially develop a metric for such comparisons, but I am not aware of this having been done.

By definition, no model-theoretic accounts can meet the QUALITATIVE desideratum. Such accounts can only compare structure to structure. Since many proponents of these views, especially partial isomorphism views, tend to be structural realists, they would simply deny that the kind of qualitative features I have discussed are relevant to scientific inquiry. But I disagree; qualitative features are frequently compared between models and targets in scientific practice and our account of the model–world relationship should be capable of capturing this fact. Similarly, isomorphism-based accounts are insensitive to CONTEXT. Since partial isomorphisms can come in degrees, proponents of such accounts could argue that contextual factors dictate the determination of when a particular partial isomorphism is good enough for the purpose at hand. But this would not be part of their account of the model–world relation, and I don't believe that they have a separate account about how to make such judgments.

These difficulties with model-theoretic accounts have lead some philosophers to look elsewhere for an account of the model–world relation. Some philosophers have urged that we reorient the question, asking about what inferences can be drawn from models, not how models represent their targets (e.g., Suárez, 2004). But a more popular response has been to search for an alternate type of relation. A number of philosophers including Cartwright (1983), Giere (1988), and Godfrey-Smith (2006), have argued that successful models are *similar* to their targets. I agree and will develop this view throughout the rest of the chapter.

8.3 SIMILARITY

Although a number of philosophers have advocated similarity as the ideal candidate for the model–world relation, it has a checkered history in philosophy and appeals to it have seemed dubious to many philosophers. For example, in 1969, W. V. O. Quine argued that any general notion of similarity is deeply problematic because we cannot explain it in terms of more empirically or logically basic notions. On this basis, he concluded that the concept of similarity is "logically repugnant" (1969, 59) and that mature sciences should and do dispense with similarity relations altogether.

While Quine's argument primarily rested on simple chemical examples, he also invoked more foundational arguments from Nelson Goodman. One of Goodman's major arguments against similarity is that appeals to it merely label something unknown, rather than giving a characterization of the relationship in question. A proper analysis ought to give a reductive definition of similarity, but, Goodman argues, no such definition could exist (1972).

Another of Goodman's arguments is that similarity is too promiscuous of a relation. For example, for any three objects, there will always be some respect in which two of the objects resemble each other more than the third. If we have a green square, a red square, and a red circle, there is no obvious pair whose members are more similar to each other than to the third.

This problem scales up beyond the trivial to real scientific examples. Consider the molecules ethyl alcohol, dimethyl ether, and diethyl ether. Ethyl alcohol has the structure: CH_3CH_2-OH; dimethyl ether: CH_3-O-CH_3; and diethyl ether: $CH_3CH_2-O-CH_2CH_3$. Which pair of molecules is more similar to each other than they are to the third? Dimethyl ether and ethyl ether are both ethers, hence they can engage in similar chemical reactions and dissolve similar substances. On the other hand, ethyl alcohol and ethyl ether both have ethyl groups (chains containing two carbon atoms). Ethyl alcohol and dimethyl ether are structural isomers, meaning that they have the same atoms, just arranged in different orders; they also have the same molecular mass. However, ethyl alcohol is completely soluble in water, whereas both ethers are only partially soluble in water. Ethyl alcohol boils at 78.4°C, while the two ethers boil at a much lower temperature (34.6°C and -23°C respectively). So depending on which features one takes to be the most salient, any pair could be judged more similar to each other than to the third.

Goodman uses this second problem to show that there can be no context-free similarity metric, either in the trivial case or in a scientifically realistic case. Such arguments led philosophers like Giere and Cartwright to restrict their accounts of model–world similarity. Giere tells us that a model must resemble its target in certain "respects and degrees" (Giere, 1988, 93), which are presumably given to us by background theory and a theorist's interests. Cartwright tells us that the relevant similarity between models and their targets is "behavioral similarity," which I interpret to mean similarity of causal structure (Cartwright, 1983).

I think that Giere and Cartwright are on the right track here. Similarity does seem to be the right kind of relationship to meet the desiderata of Section 8.1 because it comes in degrees, can be used to compare idealized models to targets, can relate qualitative features of models to targets, and so forth. However, Giere and Cartwright give us few details about what similarity supervenes on, how it depends on context, how similarity judgments are to be evaluated, and so forth. Thus, in the remaining sections of this chapter, I will introduce a novel account of the model–world relation that attempts to answer these questions. My discussion begins with the work of Amos Tversky, who developed an influential account of similarity in psychology.

8.4 ■ TVERSKY'S CONTRAST ACCOUNT

In the 1970s, Amos Tversky developed a set-theoretic account of similarity with which he tried to capture the everyday judgments of similarity and dissimilarity made by his experimental subjects. At the time, the most sophisticated theory of similarity judgments had been developed by Frank Attneave (1950) and Roger Shepard (1980, 1987), drawing on some of Quine's ideas. In Shepard's and Attneave's *geometrical* account of similarity, objects are assigned to a location in multidimensional space based on values assigned to their features. Similarity, then, is just the distance between points representing objects in this space. For example, colors might be represented as coordinates in a three-dimensional space, corresponding to their lightness, hue, and saturation. Two colors could then be compared to each other by measuring the distance between them. The closer two objects are in this feature space, the more similar they are to one another.

Tversky thought that this wasn't a fully general account of similarity for a number of reasons. For one thing, he believed that not all properties relevant to similarity judgments can be mapped onto a dimension of a property space; some features are qualitative. He also believed that not all similarity judgments were symmetric. His subjects judged that North Korea was more similar to China than China was to North Korea (Tversky, 1977; Tversky & Gati, 1978).[3] So Tversky wanted an account that was more flexible and general than the geometric account, but that could also generate the results of the geometric account when they applied.

To a first approximation, Tversky's *contrast* account of similarity says that the similarity of objects a and b depends on the features they share and the features that they do not share. His account is developed in the following way. We begin with some set of features Δ, called the feature set. These can be quantitative or qualitative and might include elements such as "is red," "is to the left of X," "will land on heads with probability 0.5," and just about anything else. For two objects a and b, we will define A as the set of features in Δ possessed by a, and B as the set of features in Δ possessed by b. Further, we choose some weighting function $f(\cdot)$, which is defined over the powerset of Δ ($\wp\Delta$). The similarity of a to b is then given by the following equation:

$$S(a,b) = \theta f(A \cap B) - \alpha f(A - B) - \beta f(B - A) \qquad (8.3)$$

For some set of features Δ, weighting function $f(\cdot)$, and term weights θ, α, and β, this equation will give us a similarity score that can be used in comparative judgments of similarity. It says that the similarity of a to b is a function of the features they share, penalized by the features that they do not share. Tversky thought that the term weights and weighting functions were context sensitive and that the rules governing them would be discovered by empirical psychology.

I think that Tversky's contrast account makes an excellent starting point for developing a formal similarity-based account of the model–world relation. Prima facie, it meets many of my desiderata for model–world relations because it allows comparisons to be made between just about any property of a model and a target. The simplest possible version of such an account would be to swap out Tversky's generic objects a and b with models (m) and targets (t). If we did this, we could write down an equation of the following form:

$$S(m,t) = \theta f(M \cap T) - \alpha f(M - T) - \beta f(T - M) \qquad (8.4)$$

where M and T are defined in the manner of A and B in Equation 8.3.

To a first approximation, I think that this the correct account of the model–world relationship. A model is similar to its target, or to a mathematical representation of its target, when it shares certain highly valued features, doesn't have

---

3. Gleitman and colleagues have criticized Tversky's claim that similarity judgments can be asymmetric. They argue that the kinds of asymmetric judgments Tversky's subjects make can be induced for sentences which are uncontroversially symmetric. A proper understanding of the semantics of Tversky's test items will explain why subjects make asymmetric linguistic judgments, even if the similarity relation is symmetric (Gleitman, Gleitman, Miller, & Ostrin, 1996). As we shall see, Tversky's formalism can be used to model both asymmetric and symmetric similarity judgments. For my purposes, nothing will ultimately turn on whether our ordinary similarity judgments are symmetric.

many highly valued features missing, and when the target doesn't have many significant features that the model lacks. Relevant features are identified in a natural or formal language and their importance is weighted relative to the goals of the scientific community.

In order to transform this basic idea into an account of the model–world relation, we need to consider in more detail where f, Δ, and the weighting coefficents come from. We also need to attend to the form of the equation itself.

8.5 ▪ ATTRIBUTES AND MECHANISMS

Let's begin with the form of the equation, which will need some initial refinements. The first refinement has to do with types of features. In an influential set of papers, Goldstone, Gentner, Markman, and Medin showed that, in everyday judgments of similarity, intrinsic features of objects are weighted less heavily than the relations possessed by objects and their parts (Goldstone, Medin, & Gentner, 1991; Gentner & Markman, 1994, 1998; Goldstone, 1994). We needn't be concerned to follow their distinction slavishly, nor their more general point that everyday similarity judgments follow structural alignment, but I think that they highlight a distinction among features that is relevant to thinking about the model–world relation.[4]

In scientific inquiry in general, it is typical to distinguish the properties and patterns of a system from the underlying mechanism that generates these properties. Because of this, I propose that we make a major division between properties and patterns on the one hand, and the underlying generating processes on the other. Call the first category *attributes* and the second category *mechanisms*. A more abstract way to think about the difference between attributes and mechanisms is that attributes are states and mechanisms are transition rules. This is very similar to Goldstone, Gentner, Markman, and Medin's distinction between attributes and relations, although my notion of mechanisms is more restricted because mechanisms are causal. Noncausal relations would be considered attributes on my account.

As an example of the distinction between attributes and mechanisms, consider equilibrium states of Schelling's segregation model. When the model comes to equilibrium, it contains racially segregated clusters and it approaches this state with a pattern of "contagion," where small clusters lead to bigger clusters. What drives these patterns are the agents' utility functions and rules for movement. Attributes such as degrees of clustering are states of the model,

---

4. Further experimental and computational work on the psychology of similarity judgments can be found in Hahn & Ramscar, 2001, Hahn, Chater, & Richardson, 2003, and Kemp, Bernstein, & Tenenbaum, 2005.

and mechanisms such as agents' movement rules are the transition rules of the model. Insofar as Schelling's model explains segregation in actual cities, then there has to be some relation between the model's attributes and the city's attributes. And there has to be some relation between the model's transition rules and the actual mechanisms that drive segregation in the city.

With this division in mind, we can take the initial account of model–target similarity described in Equation 8.4 and divide its terms for model and target features into two categories: attributes and mechanisms. I will designate these with the subscripts a and m. The expression for similarity becomes:

$$S(m,t) = \theta f(M_a \cap T_a) + \rho f(M_m \cap T_m)$$
$$- \alpha f(M_a - T_a) - \beta f(M_m - T_m) \qquad (8.5)$$
$$- \gamma f(T_a - M_a) - \delta f(T_m - M_m))$$

We now have expressions for the intersection of attributes ($M_a \cap T_a$) and mechanisms ($M_m \cap T_m$) as well as the difference between the model and target's attributes and mechanisms. Additionally, each of the six terms can now be weighted independently, which is an important part of the explanation of how theorists evaluate different kinds of models used for different purposes.

For the moment, assume that we adopt the simplest possible feature weighting function $f(\cdot)$, where each element of Δ is weighted equally. In other words, for some set A (in other words, both M and T):

$$f(A) \to |A| \qquad (8.6)$$

This equation says that each term in the similarity equation has the numerical value of its cardinality. We can also use the simplest possible weights for the individual terms. If we set $\theta = \rho = \alpha = \beta = \gamma = \delta$, we can just drop the weights from our expression. In this case, the equation becomes:

$$S(m,t) = |M_a \cap T_a| + |M_m \cap T_m|$$
$$- |M_a - T_a| - |M_m - T_m|$$
$$- |T_a - M_a| - |T_m - M_m| \qquad (8.7)$$

Equation 8.7 makes the basic structure of our modified Tversky equation clear. A model is more similar to its target when it shares more attributes and mechanisms, and is systematically penalized when the model contains extraneous detail and when it fails to capture or incorrectly captures features of the target.

Unfortunately, as written, this equation can yield similarity values that range from $-|\Delta|$ to $|\Delta|$, and hence will not be particularly informative in making comparative jugements between models because the scale itself will change with changes to Δ. It would be preferable for the scale to always be bounded in the same way so that we can compare across cases and contexts. This won't allow us to make precise comparisons across different contexts, but will let us say that, given our fidelity criteria, model m_1 and target t_1 are moderately similar, just like model m_2 and target t_2. In cases where the standards of fidelity are shared, then we should be able to say that model m_1 is more similar than model m_2 to target t.

One way to have a standardized scale would be to use a formulation that Tversky suggested in passing, but never fully developed. Rather than taking similarity to be a measure of the features shared minus the features not shared, we can take it to be the ratio of features shared to those not shared. Further, if we normalize the equation, we can ensure that similarity values are bounded between 0 and 1, corresponding to maximally disimilar and identical, relative to Δ.

Rewriting Equation 8.7 in ratio form yields:

$$S(m,t) = \frac{|M_a \cap T_a| + |M_m \cap T_m|}{|M_a \cap T_a| + |M_m \cap T_m| + |M_a - T_a| + |M_m - T_m| + |T_a - M_a| + |T_m - M_m|} \quad (8.8)$$

This equation is the core of my *weighted feature-matching* account of model–world relations.

One thing to notice about this expression is that it intuitively satisfies the MAXIMALITY constraint. When model and target share many features, S approaches 1. When they are identical, $S = 1$. When they are maximally disimilar, sharing no features at all, then $S = 0$. More generally, the limiting behavior of this expression can be written

$$\lim_{|M \cup T| \to |M \cap T|} S(m,t) = 1 \quad (8.9)$$

At this point, we can consider how the different terms would get filled for some of the examples I have discussed throughout the book. First, consider the San Francisco Bay model. We know that, within reasonable tolerance, the model is physically scaled to the Bay's features, and the tidal cycle and salinity gradient are represented accurately (attributes shared by model and target, or $M_a \cap T_a$). Mechanistically, the model has little in common with the forces producing tides (the moon) and salinity gradients (the ocean) in the Bay (mechanisms possessed by the model but not the target or $M_m - T_m$). However, the ratios of physical forces as represented, for example, by the Froude number (ratio of body's inertia to gravitational forces) are correctly represented (mechanisms possessed by the model and the target, or $M_m \cap T_m$).

The Schelling model is a very different kind of structure than the Bay model, but a similar analysis of the key terms can be made. The model is primarily aimed at generating shared attributes ($M_a \cap T_a$). It reproduces patterns of racial segregation, and choosing the correct utility function can create whatever racial exposure values were observed. On the other hand, it is almost certainly not the case that real populations have fully shared, simple utility functions like "I want at least 30% of my neighbors to be like me" ($M_m - T_m$). Similarly, even the most grid-like cities such as Philadelphia are not completely regular grids like in the Schelling model ($M_a - T_a$).

These examples should give us some idea of how the main terms of Equation 8.8 can be filled in. Now, let's turn to the issue of how we generate similarity "scores" from such an equation. This is done by the weighting function, which assigns similarity values to each term of the equation. Intuitively, what the weighting function does is to indicate the importance of various members of the feature set.

More formally, we can start by dividing our feature set Δ into two subsets corresponding to attributes (Δ_a) and mechanisms (Δ_m). The terms in Equation 8.8 correspond to elements drawn from the intersection of the powersets of these subsets. Thus, our weighting function must be defined on this intersection. In other words, $f : (\wp(\Delta_a) \cup \wp(\Delta_m)) \to \mathfrak{R}_{\geq 0}$. We further restrict f such that $f(\emptyset) = 0$. Each term must be given a non-negative weight, to keep the overall similarity score between 0 and 1. With these restrictions in place, the overall expression is then written as follows:

$$S(m,t) = \frac{\theta f(M_a \cap T_a) + \rho f(M_m \cap T_m)}{\theta f(M_a \cap T_a) + \rho f(M_m \cap T_m) + \alpha f(M_a - T_a) + \beta f(M_m - T_m) + \gamma f(T_a - M_a) + \delta f(T_m - M_m)}$$
(8.10)

This general form of the model–world relation is instantiated by many specific relations corresponding to different choices for term weights and weighting function. The remainder of this chapter will consider how Δ, f, θ, ρ, α, β, γ, and δ are chosen.

8.6 ■ FEATURE SETS, CONSTRUALS, AND TARGET SYSTEMS

In order to start filling in Equation 8.10, let's begin by considering the nature of the feature set Δ in more detail. Specifically, let us ask: What kind of terms go into Δ? Where do these terms come from? Why doesn't their presence in this account make it arbitrary and unilluminating? The simple answer to these questions is that the elements of Δ come from a combination of context, conceptualization of the target, and the theoretical goals of the scientist. Let's begin by considering what kinds of terms can go into the feature set.

Stated abstractly, the elements of Δ can be qualitative, interpreted mathematical features such as "oscillation," "oscillation with amplitude A," "the population is getting bigger and smaller," etc. They can be strictly mathematical terms such as "is a Lyapunov function." Or they can be physically interpreted terms such as "equilibrium" or "average abundance." Which of these kinds of terms should go into a particular Δ?

There is no context-free answer to this question, but part of the answer lies in the modeler's intended scope. The modeler's intended scope takes into account the research question of interest, the context of research, and the community's prior practice (Kitcher, 1993). These elements of the modeler's intended scope, in turn, determine the contents of the feature set. So ultimately the choice of scope is equivalent to the choice of Δ.

With a conceptualization of the target and model into properties in hand, the scientist can add elements to Δ. For example, an ecologist developing a Δ for the Lotka–Volterra model would include terms like "equilibrium abundance" and "maximum population size," reflecting a fairly concrete interpretation of mathematical terms.

Similar questions about what to include in Δ arise with concrete models such as the Bay model. Even though concrete properties of the model are extremely salient, both concrete and abstract properties can enter as elements in Δ. For example, dimensionless quantities such as the Reynolds and Cauchy numbers are essential for establishing the relevant degree of similarity between model and target. Hence, these numbers are included in Δ, along with concrete properties such as the physical shape of the model and the target (a more general discussion of this issue can be found in Sterrett, 2005, and Pincock, 2011).

As science progresses and more is known about a model's targets, the contents of Δ may change. Modelers might initially deem some features of models and targets important, but, as science progresses, these might be judged to be irrelevant. Similarly, new properties of targets might come to be recognized as especially important. These changes in practice and interest will occasion a change in Δ, and consequently a reevaluation of the model–world relationship. These changes alone can have the effect of rendering the model more or less similar to a target. At first, this might seem like a disadvantage of the account, suggesting that the account's flexibility precludes it giving a good analysis of the model–world relationship.

However, there are two reasons why this is not a disadvantage. First, the similarity relation that I am developing already supervenes, in part, on the modeler's construal. When context or scientific goals change, the construal will change, and aspects of the relation will change. Second, changes in the perceived quality

150 ■ Simulation and Similarity

of a model, because of its relationship to its target, are the sorts of things that happen through the history of science. Volterra believed that his model was a good way to explain the fishery dynamics of the Adriatic. Contemporary scientists judge the model to be more heuristic than predictive for any actual dynamics. At least part of this diverging judgment has to do with differences in the content of Δ, because Volterra was using a much more qualitative feature set than many contemporary ecologists.

In sum, the contents of the feature set Δ are strongly dependent on theorists' construals of their models. Their intended scope and assignment partition and individuate properties of targets and models, and those deemed relevant are included in the feature set. As this set changes, the model–world relation can change.

8.7 ■ MODELING GOALS AND WEIGHTING PARAMETERS

While the elements of Δ need to be specified with respect to specific models, targets, and contexts, a much more general account can be given about the weighting parameters for the terms in Equation 8.10. In order to do this, let me begin by addressing an ambiguity in my discussion of the model–world relation thus far. It is traditional to say that the model–world relation is the relationship in virtue of which studying a model can tell us something about the nature of a target system. But at the same time, scientists are often interested in comparing the relationship that the model actually holds to the world to the one that they are interested in achieving between the model and the world.

In isomorphism-based accounts of this relationship, the only guidance that can be given is that the model is isomorphic or it isn't. There is no way of expressing the existence of a relatively good fit between model and target, or the gradual improvement of this relationship with improvements to the model. In contrast, weighted feature-matching allows scientists to assess how close they have come to meeting their goals. It also recognizes that different goals can require different kinds of similarity relations, or at least the emphasis of different kinds of features. This is accounted for by the way in which the parameter values for each terms of Equation 8.10 are set. Drawing on my discussions of idealization in Chapter 6, let's consider several specific kinds of modeling and how they influence the weighting of the terms. For now, I will suppress the feature weighting function $f(\cdot)$, and look at expressions with this function mapped to cardinality.

The simplest case is what we can call hyperaccurate modeling, connected to the representational ideal of COMPLETENESS (Section 6.2). In this type of modeling, the theorist wants the model to contain all of the features of the target, and to neither have distortions ($M - T$) nor approximations or further abstractions ($T - M$). In this case, the theorist aims for:

$$\frac{|M_a \cap T_a| + |M_m \cap T_m|}{|M_a \cap T_a| + |M_m \cap T_m| + |M_a - T_a| + |M_m - T_m| + |T_a - M_a| + |T_m - M_m|} \to 1 \quad (8.11)$$

This equation says that our aim is for maximal similarity ($s(m,t) = 1$) when all of the mechanisms and attributes in Δ are included. The expression will be satisfied when the complement terms (e.g., $M - T$) have a cardinality of zero.

Three more interesting cases are how-possibly modeling, minimal modeling, and mechanistic modeling. In how-possibly modeling, the goal is to find some mechanism or other that can reproduce the attributes of the target. This means that the attributes of the model and target must be similar, but any plausible mechanism can be used to generate these attributes. This corresponds to $|M_a \cap T_a|$ having a high value and $|M_a - T_a|$ having a low value. We can express the goal of how-possibly modeling as:

$$\frac{|M_a \cap T_a|}{|M_a \cap T_a| + |M_a - T_a|} \to 1 \quad (8.12)$$

Another type of modeling is minimal modeling. In this type of modeling, theorists want to find one (or very few) mechanisms thought to be the first-order causal factors giving rise to the phenomenon of interest, but nothing else. A classic example of a minimalist model in the physical sciences is the Ising model (Ising, 1925), which I discussed in Section 6.1.2. When building a minimal model, the theorist wants to ensure that all of the mechanisms in her model are in the target (high $|M_m \cap T_m|$), and that her model reproduces the first-order phenomena of interest (high $|M_a \cap T_a|$), that there aren't any extraneous attributes or mechanisms in the model (low $|M_m - T_m|$ and $|M_a - T_a|$). On the other hand, the target can have all kinds of mechanisms and attributes not in the model. This corresponds to:

$$\frac{|M_m \cap T_m|}{|M_m \cap T_m| + |M_a - T_a| + |M_m - T_m|} \to 1 \quad (8.13)$$

In addition, the modeler wants to establish that:

$\exists \psi \in \Delta, \psi \in (M_m \cap T_m)$, and ψ is a first-order causal factor of T.

Finally, mechanistic modeling is the practice where theorists are interested in exploring a potential mechanism and understanding what that mechanism can produce. Sometimes, mechanistic modeling is an example of modeling without any target at all. However, when theorists have a target in mind, they are interested in generating a model which shares many mechanistic features with its target (high $|M_m \cap T_m|$), and has few mechanisms in the model not in the target (low $|M_m - T_m|$), and few mechanisms in the target not in the model (low $|T_m - M_m|$). This corresponds to trying to achieve

$$\frac{|M_m \cap T_m|}{|M_m \cap T_m| + |M_m - T_m| + |T_m - M_m|} \to 1 \qquad (8.14)$$

8.8 ■ WEIGHTING FUNCTION AND BACKGROUND THEORY

The final aspect of the weighted feature-matching relation is the weighting function $f(\cdot)$. In very general terms, the purpose of this function is to tell us the relative importance of elements and combinations of elements in Δ. While all features of models and targets that are included in the feature set are taken to have some importance in establishing similarity between model and target, some are considerably more important. The weighting function tells us the relative importance of each feature.

We can build up a general form of the weighting function by considering aspects of term weights that are common across cases. In order to satisfy Equation 8.10, we need to define $f(\cdot)$ over $\wp(\Delta_a) \cup \wp(\Delta_m)$. However, it is very unlikely that scientists could mentally represent anything remotely resembling a representation of a function being defined on this set, or even produce such a function if called on to do so. For nontrivial Δ's, there are simply too many elements in this powerset. Moreover, it is unlikely that the relative importance of features would be articulated in this way. Rather, scientists typically think about the relative weights of some or all of the elements of Δ. This means that we can substantially restrict the weighting function by requiring that:

$$f\{A\} + f\{B\} = f\{A, B\} \qquad (8.15)$$

or more formally that

$$f(X) = \sum_{x \in X} f\{x\}. \qquad (8.16)$$

This means that the total weight given on some set X will be equivalent to adding up the weight given to all of the elements of X. This would require the theorist to have access only to the weight she places on each element of Δ.

This seems more realistic, but still far from how most scientists think about the model–world relation. Even this restricted form of the weighting function assumes that scientists represent the weight of each element in Δ. But in most cases, scientists will believe that some subset of the features in Δ are especially important and might have a relative weighting of these features. We can call this subset the set of *special features*, and these will be weighted more heavily than the rest. The others will simply be equally weighted. As a default, the weighting function will return the cardinality of sets like $M_m \cap T_m$. The subset of special features will receive higher weight according to their degree of importance.

Restricting the weighting function in this way raises a new question: How do scientists determine which elements of Δ are the special features? And for those features, what weights should be put on them? In the best-case scenario, background theory will tell us about which terms require the greatest weights. For example, in the San Francisco Bay model, engineers were primarily focused on the hydrodynamic features of the model. These features were described by fluid mechanics, which could tell them that the spatial scaling of the model was not sufficient to achieve the relevant degree of model–target similarity. Their own words are instructive:

> The establishment of the similarity between model and prototype commonly makes use of laws of similitude expressed by ... the Froude, Reynolds, Weber, and Cauchy numbers which are the ratios of inertial forces to the forces of gravity, viscosity, surface tension, and elasticity, respectively. Similitude requires that each of these numbers be the same for model and prototype. In an estuary such as the Bay system, the depth, surface slope, and other features of flow are controlled by the joint effect of inertial and gravitational forces. Thus, all hydraulic quantities vary according to the Froude number. The forces of surface tension and elasticity, represented by the Weber and Cauchy numbers, do not significantly affect conditions in the Bay, so it is not necessary to simulate their effects. (Army Corps of Engineers, 1981, 6-1)

Insofar as the Corps' engineers were interested in reproducing the hydrodynamic properties of the target (T_a), their primary concern was that the Froude number for the model and the Froude number for the target were the same. This property would thus be weighted much higher than everything else in Δ.

Unfortunately, many cases of modeling in biology and the social sciences will not work this straightforwardly. In these cases, background theory will not be rich enough to make these determinations, which means that the basis for choosing and weighting special features is less clear. What happens in such cases?

When background theory is only weakly developed, the possibility of reasonable disagreement increases. There can and will be reasonable disagreements about which terms should be weighted more heavily in these cases. However, there is a sense in which choosing a weighting function is in part an empirical question. The appropriateness of any particular function can be determined using means-ends reasoning.

For example, say an ecologist decides that she is most interested in determining the dynamics of population abundance. Observational and experimental research can be organized to determine especially salient features of population dynamics, such as the ones commonly seen, or the ones that portend major changes in the population. These features are then implicitly designated the special features, and good models are the ones that can reproduce these features.

Much of the time, this kind of procedure is implicit, and becomes part of what Kitcher calls a community's *practice* (1993). When assumptions about aims and the relative importance of different aspects of models are widely shared, details about weighting functions are rarely articulated. In fact, in such cases, there may be a range of permissible weighting functions accepted by the community. But when anomalies accumulate, or different subcommunities regard models very differently, communities are forced to be more explicit about their weighting functions. Being explicit in this way can help scientists negotiate their differences.

One final point about the weighting functions: Modeling practice often involves an interaction between the development of the model and the collection of empirical data. As these practices are carried out in time, scientists can discover that they were mistaken in caring so much about particular features of their models either for empirical or theoretical reasons. My account can capture this feature of practice as changes made to the weighting function through time.

8.9 ■ SATISFYING THE DESIDERATA

Now that my account of weighted feature-matching has been outlined, we can return to the desiderata for a model–world relation and see how well the account scores. The first criterion is MAXIMALITY, that for two models a and b, $s(a,b) \leqslant s(a,a) = 1$. This is an easily proved theorem of my account (see Equation 8.9). Moreover, weighted feature-matching relations are SCALAR; unlike model-theoretic relations like isomorphism, they come in degrees.

Because the bases of these relations can be any properties of models and their targets including static patterns, dynamic patterns, monadic properties, and causal structures, my account satisfies the RICHNESS criterion. And since any of these properties may be qualitative or quantitative, it satisfies the QUALITATIVE desideratum. These desiderata are extremely difficult to satisfy on model-theoretic accounts, since these accounts only compare the actual mathematical structure of the model to the mathematical structure of a representation of a target.

Weighted feature-matching also allows models that are highly idealized to be compared to targets. This is because the account does not require wholesale matches of the structure of models to the structure of targets. It doesn't even require wholesale matches of substructures. Abstract features such as "oscillatory character" can be compared, without any further specification of the structure of such features. Thus weighted feature-matching satisfies the IDEALIZATION desideratum much more easily than any existing model-theoretic account.

The weighted feature-matching account's emphasis on the term weights and weighting functions allows for the diagnosis of extraempirical disagreements, locating the sources of disagreement in context, use, and weighting. This meets the ADJUDICATION criterion. I think that this is one of the most attractive features of the account because, not infrequently, there are extraempirical disagreements among scientists about the merit of particular models. For example, it is often the case that one party feels that it has developed a good model, but a second party, with different concerns, criticizes that model for lacking a number of important features. The analysis developed in this chapter can be of assistance in several ways. For one thing, it can help theorists focus on the specific type of modeling relation they are trying to assess. Are they trying to construct how-possibly models (Equation 8.12) or minimal models (Equation 8.13)? Once this is agreed on, theorists can discuss which kinds of weighting functions are being assumed? Is there a particular feature that one theorist sees as overwhelmingly important, while another theorist gives equal weight to a large range of features? Of course, such issues are not settled by the weighted feature-matching account, nor should they be. But the account does help to articulate the sources of disagreements and pinpoint where scientists must look to resolve differences.

Finally, weighted feature-matching reflects judgments about the relationship of models to their targets that scientists can actually make, because it draws on resources that are cognitively available: feature sets and weighting functions. In many cases, scientists' judgments about similarity can be made without making feature sets and weighting functions explicit. However, they can be made precise and explicit when needed. Weighted feature-matching thus meets the TRACTABILITY criterion.

To conclude, I would say that Cartwright and Giere are basically correct about the model–world relationship. Fruitful models are similar to their targets in certain respects and degrees. Specifically, models are similar to their targets when they share many, and do not fail to share too many, features that are thought to be salient by the scientific community. This notion of similarity begins from an everyday notion, but rejects the idea that similarity is a strictly holistic relation of resemblance. The additional structure of weights and feature sets lets us capture the similarity judgments made by scientists, who may know all the same empirical facts, but who judge the similarity of a model to its target differently.

9 Robustness Analysis and Idealization

When theorists are confronted with highly idealized models of phenomena, they require a method for determining which aspects of their models make trustworthy predictions or can reliably be used in explanations. In some cases, such as when one is modeling physical systems, fundamental theories can guide the theorist on these issues. Such theories have the resources to estimate the effect of various idealizations, providing guidance about which idealizations are acceptable when particular degrees of accuracy and precision are required. In the study of many complex systems, however, such theories are unavailable. In these cases, *robustness analysis* provides an alternative method for determining when models make trustworthy predictions about their targets.

This chapter develops an account of robustness analysis. I will begin with a reconstruction of Levins' formulation of robustness analysis along with some of Wimsatt's insights about the process. I will then give a more expanded picture of the practice, discussing the role of robust and nonrobust findings in theoretical practice. Finally, I will consider some common objections to the idea of robustness analysis, and consider the ways that robustness can assist theorists in learning about targets.

9.1 ■ LEVINS AND WIMSATT ON ROBUSTNESS

Richard Levins introduced biologists and philosophers to the notion of robustness analysis in his celebrated 1966 paper "The Strategy of Model Building in Population Biology." In this paper, Levins argues that there is a three-way tradeoff between the modeling desiderata of generality, realism, and precision. This tradeoff prevents theorists from developing single models for complex phenomena, because one model cannot be maximally general, realistic, and precise (Levins, 1966; Odenbaugh, 2003; Weisberg, 2006). Faced with a proliferation of models, how can one determine which models or parts of models are reliable? The answer, Levins argued, was robustness analysis.

According to Levins, robustness analysis can show "whether a result depends on the essentials of the model or on the details of the simplifying assumptions" (1966, 20). It allows us to learn if a model's result is merely an artifact of an idealization, or if it is connected to a core feature of the model. We can do this by studying a number of similar but distinct models of the same phenomenon.

> [I]f these models, despite their different assumptions, lead to similar results, we have what we can call a robust theorem that is relatively free of the details of the model. Hence, our truth is the intersection of independent lies. (Levins, 1966, 20)

Following up on Levins' ideas, Wimsatt explains:

> [A]ll the variants and uses of robustness have a common theme in the distinguishing of the real from the illusory; the reliable from the unreliable; the objective from the subjective; the object of focus from artifacts of perspective; and, in general, that which is regarded as ontologically and epistemologically trustworthy and valuable from that which is unreliable, ungeneralizable, worthless, and fleeting. (Wimsatt, 1981, 128)

From Wimsatt, we learn that the aim of robustness analysis is to separate the scientifically important parts and predictions of our models from the illusory ones which are accidents of representations.[1] These reliable parts are what Levins called *robust theorems*.

Levins never makes the procedure of robustness analysis explicit, but he illustrates it with an example of what he believes to be a robust theorem:

> In an uncertain environment species will evolve broad niches and tend toward polymorphism. (Levins, 1966, 20)

This result, he tells us, can be derived from three kinds of models: the fitness set model (Levins, 1962), a model using the calculus of variation, and a genetic model (Levins & MacArthur, 1966).[2]

The implied procedure behind this example seems to be the following: First, identify a series of models for a target phenomenon. Second, look for a prediction common to these models. When you find the common prediction, then you have found your robust theorem and you can treat this theorem as well confirmed. More formally, for some set of models $M \in \{m_1, m_2, \ldots, m_n\}$, if $(\forall m_i \in M) m_i \models T$, then T is true.

Orzack and Sober (1993; see also Orzack, 2005) give this reading of Levins and point out that the procedure outlined above will almost always lead to an invalid inference. It would only generate a valid inference if M is an exhaustive set of all possible models. Thus, they conclude, robustness analysis adds very little to scientific inquiry.

I reject this conclusion because I don't think that this is an adequate reconstruction of Levins' ideas about robustness. Part of the problem is that Orzack

---

1. Wimsatt also thinks that robustness analysis has an empirical side, which can demonstrate when some phenomenon does not depend on particular parts of an empirical system. See Wimsatt, 1981, and Wimsatt, 1980, for details.
2. Subsequent research, for example Seger & Brockmann (1987), has called into question whether this particular phenomenon is actually robust.

and Sober have interpreted the term 'robust theorem' too literally. In their analysis, T is literally a theorem of the models in M. But this is unlikely to be what Levins had in mind. In the next section, I will argue that a more careful definition of 'robust theorem' will illustrate the value of robustness analysis.

9.2 ■ FINDING ROBUST THEOREMS

Like Levins, I think that the core of robustness analysis is the search for robust theorems. This search can be thought of as a two-step procedure. The procedure begins by examining a group of models to determine if they all predict a common result, the robust property. The second step involves analyzing the models for the common structure which generates the robust property. Results from the first and second steps are combined to formulate the robust theorem itself, a conditional statement linking common structure to robust property, prefaced by a *ceteris paribus* clause. Let's look at the procedure in more detail.

In the first step, theorists examine a group of similar but distinct models, looking for a robust behavior. During this stage, it is important that they collect a sufficiently diverse set of models so that the discovery of a robust property does not depend in an arbitrary way on the set of models analyzed.

For example, say that we found the Schelling segregation result especially striking. We know that the model from which this behavior is generated is highly idealized with respect to any real city. So we construct a number of related models, varying the kinds of idealizations that we make. We could vary the regularity of the grid, the definition of a neighborhood, the number of attributes the agents care about, the heterogeneity of the utility function, the form of the utility function, the complexity of the decision procedure, and so forth. This would generate a number of similar, but distinct models, and we would examine these models to see if they also displayed the characteristic pattern of Schelling segregation.[3]

The first step is either followed by or conducted in parallel with the second, which involves finding the core structure which gives rise to the robust property. In straightforward cases, the common structure is literally the same physical, mathematical, or computational structure in each model. In such cases, one can isolate the common structure and, using the kinds of analyses I discussed in Section 5.2, verify the fact that the common structure gives rise to the robust property. However, such a procedure is not always possible because models may be developed in different computational or mathematical frameworks, or may represent a similar causal structure in different ways or at different levels of

---

3. In fact, Schelling's pattern of segregation is even more robust then even Schelling thought. See Muldoon, Smith, and Weisberg (2012) for details.

abstraction. Such cases are much harder to describe in general, since they rely on theorists' abilities to judge relevantly similar structures. In the most rigorous cases, theorists can demonstrate that each token of the common structure gives rise to the robust behavior and that the tokens of the common structure contain important mathematical similarities, not just intuitive qualitative similarities. However, there are occasions where theorists rely on judgment and experience, not mathematics or simulation, to make such determinations.

The results of the first two stages of robustness analysis can be combined to formulate robust theorems. They have the following general form:

> *Ceteris paribus*, if [common causal structure] obtains, then [robust property] will obtain.

For example, Volterra's discovery that general biocides increase the relative proportion of the prey can be formulated as follows:

> *Ceteris paribus*, if a two-species, predator–prey system is negatively coupled, then a general biocide will increase the abundance of the prey and decrease the abundance of predators.

Once a theorist formulates a robust theorem, the final part of robustness analysis is to try to determine the extent of the theorem's robustness. Obviously every model cannot be investigated, but there are a number of procedures for casting a wide net of investigation.

9.3 ■ THREE KINDS OF ROBUSTNESS

Very few robust theorems obtain universally. Defeating conditions, the sort finessed by appending the *ceteris paribus* clause to the beginning of the theorem, are always possible. So robustness analysis must also involve an investigation of the limits of robustness, the conditions under which a model will stop generating the robust property. When this sort of analysis is carried out as broadly as possible, it may be possible to replace a robust theorem's general *ceteris paribus* clause with a very specific statement of the conditions which defeat the efficacy of the core structure in generating the robust properties.

One way to determine if the common structure is being instantiated and if any preempting causes are present is an empirical investigation. While this is the most reliable way to ensure that a robust theorem can be applied, it is often impractical or impossible to collect the relevant data. Fortunately, there is an alternative which, while not completely reliable, can give us good reasons to believe the predictions and explanations of robust theorems. The alternative involves answering two key questions:

1. How frequently is the common structure instantiated in the relevant kind of system?
2. What defeats the core structure giving rise to the robust property?

Although the first question is best settled empirically, it can be partially addressed using techniques associated with robustness analysis. The key is to ensure that a sufficiently heterogeneous set of situations is covered in the set of models subjected to robustness analysis. If a sufficiently heterogeneous set of models for a phenomenon all have the common structure, then it is very likely that the real-world phenomenon has a corresponding causal structure. This would allow us to infer that, when we observe the robust property in a real system, then it is likely that the core structure is present and giving rise to the property.

The second question can be more easily addressed with robustness analysis itself. In particular, three different kinds of robustness analysis—*parameter robustness, structural robustness,* and *representational robustness*—investigate different ways that the core structure can be defeated. Parameter robustness analysis involves examining what happens when the values for the model description's parameters are varied. Structural robustness analysis involves adding new mechanistic features to the model. And representational robustness involves re-representing the mechanistic features of the model in a new representational framework. The remainder of this section will explore these three types of robustness analysis.

9.3.1 Parameter Robustness

Parameter robustness involves an examination of a property's robustness to changes to the model description's parameters. The basic idea is to determine, within some range, if changes to a model description's parameters will change the behavior of the model, specifically the behavior of an alleged robust property. With each change of the parameter values, we technically generate a different model, but these are all within the same family. So most theorists regard this as if they were investigating a single model.

In some cases, a robust property can be demonstrated algebraically. This is the case for Volterra's demonstration of the Volterra property, that general biocides increase the relative number of prey. But in many cases, parameter values must be investigated by choosing a value and computing the result of the model. This sounds straightforward, but sometimes it is not.

When models have relatively few parameters, and those parameters can take their values only from a small range, lest they lose all empirical meaning, parameter robustness is easy to investigate. One simply programs a computer to "sweep" through the values, perhaps doing multiple repetitions of each value of the model if it has stochastic elements. However, as the number of parameters increases, and as the range of reasonable values for these parameters increases, it quickly becomes computationally expensive to sweep through all parameter values. In such cases, theorists look to other methods.

There are many different tricks of the trade here, but they basically boil down to three strategies. The first is to sample the parameter space. The second is to start from the region where the robust property is found, and move away from this point in parameter space, looking to see where, if anywhere, the property is no longer exhibited. The third strategy is an active search of the space, looking for a place in the parameter space where the property doesn't exist. The success of this strategy relies on there being some gradient of approach towards the region of interest.

9.3.2 Structural Robustness

The second kind of robustness analysis is structural robustness analysis. In this kind of analysis, the theorist considers changes to the model's mechanistic attributes. For concrete models, this is done by physically altering the model. For mathematical models, new mathematical structure is introduced, typically by adding additional terms or couplings between terms to the model description. Computational models are altered by changing the model's procedure, possibly by including new state variables.

A fairly complete case of structural robustness has been the many followup analyses of the Lotka–Volterra model.[4] These studies typically start with Volterra's original model, but add in new terms representing predator satiation, the ability of prey to seek cover, multiple sources of food for the predator, or even complex adaptive behaviors such as learning. In principle, any ecological interaction could be added to the model.

Historically, it was structural robustness analysis that convinced theorists that the Volterra principle is true, but that many other properties of the Lotka–Volterra model are not robust. For example, both neutral stability and undamped oscillations are properties of the Lotka–Volterra model. However, adding any density dependence at all (i.e., a maximum number of organisms that can survive in the environment no matter how large) destroys both of these properties. On the other hand, the Volterra property is robust against this change to the model's structure.

---

4. For a comprehensive review of the classical literature about predation models, see Royama, 1971. For more contemporary discussions including the history of predator–prey modeling, see Berryman (1992), Hanski et al. (2001), Briggs and Hoopes (2004), and Jurrell (2005).

Structural robustness analysis allows the theorist to address a different set of uncertainties than the ones addressed with parameter robustness. Using this approach, the theorist can probe which parts of the causal structure represented by her model are really essential for the production of an observed behavior of the model.[5] This can take the form of adding new components to the causal structure, where one starts from a very minimal model such as the Lotka–Volterra model and adds new causal interactions. It can also involve removing factors, such as when one starts with a complex model, calibrated to a particular system, and removes factors to see which ones really make a difference. Structural robustness thus helps theorists isolate which components are necessary for the production of an important property, and which ones produce less robust properties.

9.3.3 Representational Robustness

The third kind of robustness is called representational robustness. Whereas parameter and structural robustness analyses vary the mechanistic attributes of the model to see how the emergence of the phenomenon is sensitive to these attributes, representational robustness analysis lets us probe a different feature of our models. It holds these attributes fixed, but analyzes whether or not the way these attributes are represented make a difference to the production of a property of interest.

While there are ways of ascertaining representational robustness within each of the three main types of models, the most dramatic kind of representational robustness is when the type of model itself is changed. This would be the case, for example, when a concrete model is rendered mathematically, or a mathematical model rendered computationally.

I know of no general guidelines for this type of robustness analysis, but can give some examples. The first was an early attempt by the Corps of Engineers to construct an electrically implemented analogue model of the San Francisco Bay (Harder, 1957). In their extensive report on model-based tests of the Reber and other plans to add saltwater barriers to the bay, the Corps reported on the representational robustness of their model. They noted that the

> ... increases in [barrier-free] tidal ranges from the hydraulic model ... were from 0.9 to 1.2 feet less than corresponding increases in ranges from the electric analog [...]. However, a comparison [of the two models with the barriers in place] shows that the actual experimental tidal ranges with the barrier in place are in closer agreement. (Army Corps of Engineers, 1963b, 82)

---

5. This aspect of robustness analysis is discussed especially clearly by Odenbaugh (in press).

The Corps, of course, had very good reason to believe the results of their hydraulic model; it was extremely well calibrated to its target. However, they did perform this bit of robustness analysis, showing that the model was only partially representationally robust. And they used the nonrepresentational robustness of their model to raise a cautionary note. They asked:

> Which model is correct and what causes the difference?
> This question can probably not be answered by anybody at this time, but some quite important reasons may be given for such deviations under the particular circumstances.
> ... In order to duplicate the flow conditions in the [Bay and adjacent waterways], it is necessary to first accurately describe the waterway in every aspect that may have an influence on the flows in it. Such a description does not only include the overall geometry of the waterway, but also its roughness, which is one of the factors that is very difficult to observe. The roughness and other qualities of the surfaces can never be sufficiently well described by geometry alone because this geometry is much too complicated in detail to predict the resulting resistance to a flow. (Army Corps of Engineers, 1963b, E-1)

After describing the many reasons that this procedure is especially complicated for the San Francisco Bay and Delta system, the engineers concluded with the following remark:

> From all of the above, it can be seen that to achieve a perfect verification is virtually impossible, and that even a satisfactory verification requires extreme care and attention to both practical and theoretical matters. (E-3)

In other words, the lack of representational robustness here is probably the result of the extreme dependence of the model's dynamics to small details of the geometry and roughness of the model's bottom surface. Nevertheless, even though the models do not predict exactly the same thing, they are quite close analogues.

A second example of representational robustness, and one that is much more common, is the re-expression of a classical mathematical model in a computational framework. As an example, consider an individual-based computational analogue of the Lotka–Volterra model that I developed with Ken Reisman.

Classical ecological models are population-level: They contain no explicit representation of individuals or their properties, only the statistical aggregates of those properties. In contrast, individual-based models (IBMs) explicitly represent individuals and their properties. An IBM includes a set of state variables for each individual within the model population. It also includes assumptions about how individual organisms in the population behave, develop, and interact over time. Since IBMs often contain thousands of variables, their dynamic consequences are usually investigated via computational simulation rather than mathematical analysis.

In our IBM version of the Lotka–Volterra model, we assume that individuals move about on a 30 × 30 toroidal lattice composed of 900 cells. Each individual has three variables: a binary variable denoting whether the individual is predator or prey, and two integer variables denoting a vertical and horizontal position on the lattice. Time is discrete; a global clock advances one tick at a time. For each tick of the global clock, all individuals execute a fixed set of rules that determine how they move on the lattice, reproduce, die, and interact with others. The rules for predators are as follows:

Movement rule: Move one step in a random direction.
Predation rule: Check if there are any prey on the current cell. If so, select one at random, catch it, and pick a random number from 1 to 100. If this number is less than or equal to the parameter *predator-conversion*, then reproduce.
Death rule: Pick a random number from 1 to 100. If this number is less than or equal to the parameter *predator-death probability*, then die.

These rules, when executed by each predator on the lattice, correspond roughly to assumptions made in the Lotka–Volterra model. But notice that these rules are not determined by that model. To translate any population-based model into individual-based terms, we must make explicit assumptions about individuals that were either implicit or undefined in the population-based version. This means that there is typically no uniquely correct way to carry out the translation from population-based to individual-based models.

The IBM that we designed assumes that predators move randomly on a two-dimensional toroidal lattice. The Lotka–Volterra model, on the other hand, makes no assumption about movement at all. It is consistent with the assumptions that all individuals move, that some individuals move, or even, strictly speaking, that no individuals move. It places no explicit constraints on what intrinsic or environmental factors determine movement or even whether the predator and prey move in a probabilistic or deterministic fashion.

Moreover, this IBM assumes that predators catch prey by randomly selecting one prey individual from all that are located on the same cell. Once again, this is one of the many possible assumptions we could have made to develop an IBM analogue of the Lotka–Volterra model. We could have represented predation without using a spatial lattice, where predators randomly choose prey individuals from the whole prey population. We could also have used a different predation rule on a lattice. For example, the predation rule could have stated "if a predator is within one cell of a prey, then the prey is consumed." The Lotka–Volterra model does not strictly correspond to any of these particular assumptions. A modeler who wishes to construct an IBM, however, must make an explicit decision about them.

The rules for the prey are as follows:

Movement rule: Move one step in a random direction.
Reproduction rule: Pick a random number from 1 to 100. If this number is less than or equal to the parameter *prey-reproduction probability* then reproduce.
Death rule: Check if I have been caught by a predator. If so, then die.

Together, the predator and prey rule-sets comprise one possible IBM interpretation of the Lotka–Volterra model.

I have already mentioned that the Volterra principle is structurally robust, that is, it sustains major changes to its causal structure described by the Lotka–Volterra model. What happens when we change to this individual-based version? Does the principle still hold? The answer, surprisingly, is no, but this is because of a technicality. Regardless of the parameters, one or both species in this model invariably goes extinct. Since the IBM does not exhibit coexistence of species, even in the short or medium term, it cannot exhibit the Volterra property. To test the representational robustness of this property and the Volterra principle itself, we must begin with a predator–prey model which has quasistable behavior for a reasonable length of time. Thus Reisman and I constructed a second IBM analogue of the Lotka–Volterra model, hoping to find a way to have stable oscillations.

Our second Lotka–Volterra IBM ensured stabilized oscillations by adding density dependence. For this model, we assumed that the size of the prey population is limited by availability of food in the environment (for convenience, we will call the food "foliage," but it could represent any naturally available resource). We assumed that each cell of the lattice either contains a unit of foliage or not. When eaten by a prey individual, the unit of foliage disappears, and it then has a certain probability of reappearing at any subsequent tick. The revised rules for the prey are as follows:

Movement rule: Move one step in a random direction.
Foraging rule: Check if there is foliage on the current cell. If so, eat it, and pick a random number from 1 to 100. If this number is less than or equal to the parameter *prey-conversion probability*, then reproduce.
Death rule: Check if I have been caught by a predator. If so, then die.

The rule-set for predators remains the same.

After investigating many different initial states and parameter sets, we concluded that there is a wide range of parameter conditions for which this model exhibits oscillations in the numbers of predators and prey for very long periods of time.

To check whether the model exhibits the Volterra property, we simulated the effect of a general biocide that would elevate the death rate of both the predators and of the prey. Fortunately, the individual-based framework makes it easy to simulate the dispersion of a general biocide into our model system. We performed the following perturbation. We began with a typical simulation of the predator–prey system and waited long enough for the temporal average size of each population to reach a steady state. Next, we asked the computer to randomly select some cells on the lattice to become "poisonous," so that any predator or prey that lands on the cell will die. Since movement is random, predators and prey are equally likely to die as a result of landing on poisonous cells, and the result is an increase in the death rate of both populations. Finally, after waiting for the average size of each population to reach a new equilibrium, we measured the change to the population sizes.

After performing this perturbation over a broad range of parameter settings, we found that introduction of a general biocide tended to increase the average size of the prey population and to decrease the average size of the predator population (Figure 9.1). Thus, this revised IBM does exhibit the Volterra property. Moreover, since this model also defines a negatively coupled predator–prey system, it satisfies the Volterra principle. In other words, despite moving from a population-based to an individual-based framework, and despite altering various assumptions of the Lotka–Volterra model, the Volterra principle still held. This suggests that the Volterra principle is representationally robust.

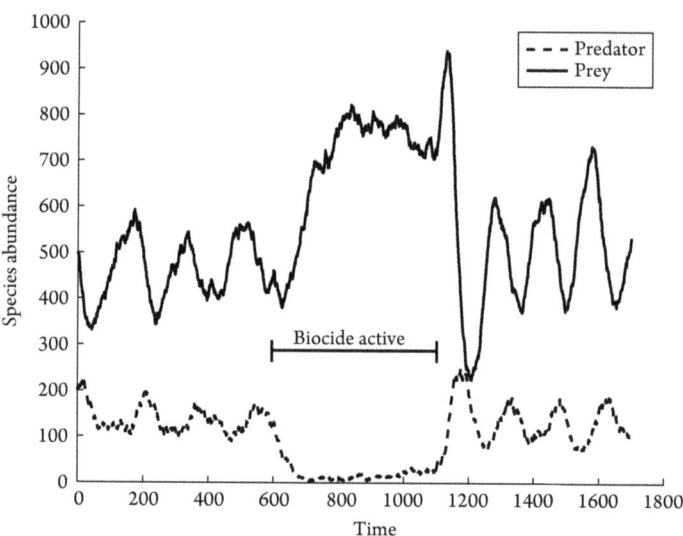

Figure 9.1 Output of Weisberg and Reisman's density-dependent individual-based predation model with biocide perturbation.

9.4 ROBUSTNESS AND CONFIRMATION

With a more complete account of robustness analysis articulated, I want to return to what seems to have motivated Orzack and Sober's original criticism of Levins. They interpreted Levins as offering a nonempirical form of confirmation via robustness analysis. They take Levins' argument that robust theorems are the "truth at the intersection of independent lies" to be a claim about some degree of confirmation for these theorems. However, Orzack and Sober reject this. Robustness analysis is a set of procedures for manipulating models, and since only observations and experiments can provide confirmation, robustness analysis cannot (Orzack & Sober, 1993; see also Forber, 2010). In light of these critiques, does robustness analysis offer any kind of confirmation to robust theorems?

I believe that the answer is no. Robustness analysis does not itself bestow confirmation on robust theorems for the reasons given above; only empirical data can confirm. However, robustness analysis can play a role in confirmation. In the remainder of this chapter, I will try to explain this role.

In order to understand robustness analysis's confirmation-theoretic role more clearly, let's step away from robustness analysis and ask a more general question: Does modeling itself ever play a role in confirmation? I think that the answer is clearly yes. Say that you have a model which has been established to have high fidelity with respect to its target. As an example, consider the San Francisco Bay model, which was shown to have high fidelity not only for the dates to which it was calibrated, but for the velocities, tides, and salinities of the Bay more generally.

Now say that we want to learn something new about the Bay. What would happen, for example, if we put a new sewer outfall near the San Mateo bridge? The only way to be sure what would happen, of course, is to build the sewer. But the advantage of having a high-fidelity model is that we can simulate the sewer outfall in the model and watch what happens. Say that the model showed us that a 1 ppm level of contamination introduced at the middle of the San Mateo bridge takes five tidal cycles to flush. We would take ourselves to be justified in believing:

> If 1 ppm of contamination is introduced in the middle of the (real) San Mateo bridge, then the Bay will take five tidal cycles to flush the contamination to undetectable levels.

And if that seems a bit too strong, we certainly seem justified in believing something like this:

> *Ceteris paribus*, if low-level contamination is introduced in the middle of the (real) San Mateo bridge, then the Bay will flush the contamination to undetectable levels within a few days.

We are justified to believe this statement because the model has been shown to have high fidelity with respect to its target, and this particular behavior is within the realm of features that the model has been shown to represent with high fidelity.

Let's look a little more closely at what role our hypothetical analysis of the Bay model played in confirming this theorem. The key step was the manipulation and analysis of the model. This manipulation showed us a property that the model possessed, but that we didn't already know about. The reason that this knowledge about the model could be transferred onto the Bay is because of the established similarity between model and Bay. So there was really a suppressed premise in the claim above. Made explicit, the claim would read:

> *Ceteris paribus*, if the hydrodynamics of the San Francisco Bay are similar to the ones in the model, then if low-level contamination is introduced in the middle of the (real) San Mateo bridge, the (real) Bay will flush the contamination to undetectable levels within a few days.

Despite seeming well calibrated in general, the model was never evaluated for being able to reproduce containment outfalls in the middle of the San Mateo bridge at a concentration of 1 ppm. Why, then, should we believe that the relevant results of manipulations to the model will result in the same behavior in the target? We should believe them because fluid dynamics can tell us that, when model–target similarity is achieved, this kind of manipulation will not disturb the model–target similitude. The similitude established in general will transfer to the case of interest. So it is a general background fact, confirmed independently of the particular model and particular question, that carries the confirmation-theoretic burden in such a case.

Now let's return to robustness analysis. Robustness analysis is generally employed when models have many idealizations present. This precludes the easy move from analysis of a high-fidelity model to an analytical result that we are licensed to accept. How, then, can the procedure of robustness analysis take the place of having a single, high-fidelity model?

We certainly should not infer from the demonstration of a robust theorem to simple statements about the ubiquity of robust properties. For example, even a wide-scoped robustness analysis for Schelling-like models of segregation doesn't license us to believe that:

> Segregation is inevitable.

Instead, robustness analysis generates a robust theorem with the following form:

> *Ceteris paribus*, if agents' decisions about where to live are guided by the Schelling utility function and movement rules, then segregation is inevitable.

I think that we can be quite confident in asserting this claim. But why? The key is to see what specifically is being claimed. The theorem makes no claim about the frequency with which the robust property obtains in real-world targets; rather, it makes a conditional claim about what happens when the Schelling utility function and movement rules are instantiated in a target.

In order to establish the truth of such a conditional, we need to show that this logical consequence of our model is mirrored as a causal consequence in the world. The data needed to confirm this inference do not show that the model is similar to its target. Rather, these data are very general facts about the representational capacity of the framework in which the models are embedded. Specifically, theorists must have data that demonstrate the adequacy of the underlying modeling framework for representing the model's causal consequences. I call this demonstration of representational capacity *low-level confirmation*.

Theorists must establish low-level confirmation in every scientific domain, showing that the frameworks in which their models and theories are framed can adequately represent their phenomena of interest. By way of example, consider models of population growth. Standard issues in confirmation theory concern whether a particular kind of model, such as the logistic-growth model, is confirmed by the available data. However, there is a prior confirmation-theoretic question that is often asked only implicitly: If the population *is* growing logistically, can the mathematics of the logistic-growth model adequately represent this growth?

Theorists rarely articulate questions about low-level confirmation explicitly in research articles. However, using models to make predications and generate explanations requires that the relevant model structures have the representational capacity to represent empirical facts about the world. This has to be established in some way. In some cases, it is established implicitly, by demonstrating that the relevant computational or mathematical structures have a good track record and have been successfully deployed to make correct predictions in the past. This is certainly true of the use of algebra and calculus in many scientific areas. But the representational capacity of these structures may also be investigated explicitly by mathematicians, such as when a new kind of mathematics is introduced, or an old mathematical structure is deployed in a new area.

Robustness analysis also relies on low-level confirmation, insofar as it can identify theorems that are well confirmed. Since low-level confirmation is what ensures that a logical consequence of a model is mirrored as a causal consequence in the world, it is what bequeaths confirmation to robust theorems. By demonstrating that Schelling-like utility functions lead to segregation and that Schelling-like models can be used to represent targets, we are warranted in accepting the robust theorem above.

Thus, robustness analysis is not a nonempirical form of confirmation as Orzack and Sober suggest. It does not confirm robust theorems. Rather, it identifies hypotheses whose confirmation derives from the low-level confirmation of the mathematical framework in which they are embedded. And it can also help scientists discover the situations where they can expect the robust theorem to be defeated.

10 Conclusion: The Practice of Modeling

Not long before his death in 1960, John Reber mailed one of his final missives, titled "Our Perpetual Gift to California and the Nation: A Master Plan for the Vast San Francisco Region" (Reber, 1959). This document recounts how he began planning his Bay project in 1907, and spent the first quarter of the 20th century gaining "incomparable knowledge of California and its people." He also writes that he spent every day of the second quarter of the 20th century, from 1932 to 1957, "educating the people anent the Reber Plan via thousands of lectures, conferences, letters, articles, radio and TV." Because of his efforts, the government allocated more than $1.5 million (over $12 million in today's dollars) to studying his grand plan.

Ironically, Reber's zeal carried within it the seeds of its own destruction. Through the construction and analysis of the Bay model, the culmination of Reber's decades-long effort, the Army Corps saved the Bay area from the devastation that Reber's perpetual gift would have caused.

The Corps' analysis of the Reber plan is a paradigm case of modeling: The indirect representation and analysis of a target system using a model. This book has developed an account of this practice, as well as accounts of models and model–world relations needed to understand how this practice works. In this concluding chapter, I want to recapitulate the main aspects of my account and show how they give us a full picture of the practice of modeling. I begin by asking: What is a model?

Models are potential representations of target systems. They come in three types: concrete, mathematical, and computational. Each type of model consists of two parts. The first part is the model's *structure*. The main way in which concrete, mathematical, and computational models differ is in their structures. Concrete models are constructed out of physically existing structures, mathematical models out of mathematical structures, and computational models out of procedures. These different kinds of structures have differing *representational capacities*.

The second part of the model is its *construal*, which is a modeler's interpretation of the structure. Construals are themselves composed of three parts: *Assignments* give a denotation to each structural element of the model. For example, an assignment can take a set of points in a space (the structure corresponding to a set of differential equations) and allow them to denote physical quantities. Points thus become quantities such as population abundances, and trajectories become the changes in those quantities through time. Closely connected to the assignment is a modeler's *intended scope*. This specifies which aspects of potential target phenomena are intended to be represented by the model, and which parts are simply excess structure.

Fidelity criteria are the final part of a modeler's construal. These criteria are the standards that theorists use to evaluate a model's ability to represent real phenomena. This is an absolutely crucial part of my picture because, as my examples have shown, the same exact model can be judged as being better or worse depending on theorists' evaluative standards. If one community sees a particular model as merely giving a how-possibly explanation, then a highly idealized model might be seen as making an important contribution, despite leaving many details of a real phenomenon unexplained. Another community, however, might require extremely accurate and precise predictions. In such a case, the same model might be seen as a poor simulacrum.

Models stand in two distinct relations. They are *specified* by *model descriptions* and they are potential representations of *target systems*. Model descriptions are mathematical, verbal, computational, or even pictorial descriptions of models. In the case of mathematical models, manipulation and analysis of the model are conducted by manipulation and analysis of the model description. Concrete and computational models can be manipulated more directly. There is a very tight link between models and model descriptions. All logically consistent model descriptions specify at least one model, and all models *satisfy* at least one model description. Vagueness and imprecision in model descriptions mean that those descriptions will be satisfied by families of models as opposed to single models.

The second relation is between models and target systems, which are parts of real-world phenomena. To generate a target, theorists choose some phenomenon in the world that they wish to study. From the full contents of the phenomenon, they *abstract*, omitting all but the relevant features of this phenomenon. This process generates the target system. Volterra's phenomenon, for example, was the Adriatic fisheries after World War I. His target system consisted of the abundance of certain kinds of fish, fishery practices, and other causal factors driving the population dynamics.

Simple cases of modeling have a single model standing in relation to a single target. But such cases are uncommon. More commonly, modelers intend their models to represent generalized targets, such as sexual reproduction in general or covalent bonding in general. This requires that their models be similar to a family of related targets. Models can also be used to study nonexistent targets such as perpetual-motion machines. Although we can say that models are similar to these nonexistent targets, investigating such models often tells us about actually existing targets.

I have now characterized the relata, models and targets, but what is the relation? In this book, I have argued that model–target relations are similarity relations, and I have offered a *weighted feature-matching* account of similarity. This account says, roughly, that models stand in representational relations to their targets in virtue of sharing some set of highly important features with their targets, not lacking too many of these features, and not having too many extra features. Scientific context, as specified by the modeler's construal, determines the choice and weighting of important features.

Models are not always related to a single target, which means that this simple picture of the model–world relation needs to be made more complex. One of the advantages of the weighted feature-matching account of similarity is that it can be "tuned" to the kinds of modeling intentions found in differing modeling practices.

For example, when a theorist wants to produce a *how-possibly explanation*, then she wants to find any plausible mechanism that can produce the target phenomenon. In the language of the weighted feature-matching account, this corresponds to $|M_a \cap T_a|$ having a high value and $|M_a - T_a|$ having a low value. The theorist thus wants to maximize the following expression:

$$\frac{|M_a \cap T_a|}{|M_a \cap T_a| + |M_a - T_a|} \tag{10.1}$$

When this expression equals one, then the model and the target share all of the attributes in the modeler's feature set.

The weighted feature-matching account has the resources to explain the relationship between all three types of models and their targets. Features of concrete models are directly compared to features of targets. For mathematical and computational models, features of models are compared to mathematical representations of features of targets. Moreover, my account could be used to compare thought experiments or mental models to their targets because thought experiments and targets can share features.

Since my account of the model–target relationship is dependent on theorists' intentions or construals, disagreements about the model–target fit are possible. My account allows for this possibility, as indeed it needs to in order to correctly characterize scientific practice. But my account also presents resources for identifying and analyzing disagreements, because it allows theorists to explicitly characterize the standard against which they are measuring their model. Are they trying to construct how-possibly models, for example, or minimal models? Once they reach agreement, or at least understanding, about their goals, they can further pinpoint any disagreement about the weighting functions that they are implicitly assuming. Is a particular feature treated as most important by one theorist, while the other the theorist gives equal weight to a large range of features? Such issues are not resolved by my account because they depend on scientific goals, background knowledge, and background theories, not the form of the similarity relation. What my account can do is to help to locate sources of disagreement and to give scientists a way to explicitly formulate their standards of fidelity.

Summing up, models consist of structures and construals. They are related to model descriptions via relations of specification, and they are related to target systems via relations of similarity. But even when theorists aim to build models that represent targets in virtue of their similarity, they are not always aiming to make the models maximally similar to their targets. Theorists often intentionally introduce *idealization* to their models.

Models are idealized when they are intentionally distorted relative to their targets. I understand idealization as an activity guided by a set of norms, which I have called *representational ideals*. Different types of idealization are thus guided by different norms. *Galilean idealization* is the process of introducing distortions into models with the goal of simplifying. Such distortions will be removed as more data are collected and computational techniques are improved. *Minimalist idealization*, on the other hand, involves distorting so as to isolate the most important causal factors. The models generated by minimalist idealization are prized for their explanatory power. Finally, when dealing with highly complex systems, modelers may construct multiple models for their targets, each capturing different aspects of these targets. The models generated by *multiple-models idealization* each make distinct claims about the nature and causal structure giving rise to a phenomenon.

The practice of idealization, and especially multiple-models idealization, means that theorists will often generate a proliferation of models of the same target. Managing this proliferation requires some way of identifying which aspects of models make trustworthy predictions and can reliably be used in explanations. I have argued that one way for theorists to get a handle on this proliferation is through the use of *robustness analysis*. In this analysis, theorists compare the models they have generated to closely related models in order to see what the models agree about. From these comparisons among models, they can generate *robust theorems*, conditional statements that identify the connection between a core causal mechanism and some property of the models. Robustness analysis does not confirm robust theorems. Rather, it identifies theorems whose confirmation derives from the low-level confirmation of the mathematical framework in which they are embedded. It can also be used to find situations where we should expect the phenomenon described by the robust theorem to be defeated.

I have now reviewed what models are and how they are related to their descriptions and their targets. The final issue in this book concerns the use of models. Put simply, models are constructed and aimed at targets in order to learn about these targets. Theorists can indirectly analyze targets by analyzing their models.

This analysis takes many different forms depending on the type of model, the interests of the scientist, and pragmatic factors including time, available computational power, and so forth. Sometimes, the goal is a complete analysis of the model, where the theorist will learn about all of the model's static and dynamic properties, allowable states, transitions between states, factors that initiate transitions between states, and dependence of states and transitions on one another. In other cases, specific goals will dictate which subset of the features of the model is studied. There are no completely general recipes for analyzing models except in the simplest cases. However, many classes of techniques can be deployed as appropriate, including analyses which give algebraic solutions, numerical techniques which give approximate solutions, simulations involving complete or probabilistic sampling of model behaviors, and, in the case of concrete models, literal experimentation.

My account of modeling has been guided by two main principles. The first is that the way modeling is actually practiced should inform our accounts about the nature of models and of model–world relations. The second guiding principle is that there is a wide range of modeling practices, including the construction of a single highly accurate representation of a single target and the modeling of nonexistent targets. My application of these principles has led me to consider the similarities and differences among examples of concrete, mathematical, and computational modeling. The account that I have developed may not always be simple and tidy, but I believe that it can help us make sense of this important and distinctive theoretical practice.

■ REFERENCES

Ankeny, R. A. (2001). Model organisms as models: Understanding the "lingua franca" of the human genome project. *Philosophy of Science, 68*(3), S251–S261.
Army Corps of Engineers. (1963a). *Technical report on barriers: A part of the comprehensive survey of San Francisco Bay and tributaries, California (Main report)*. San Francisco: Army Corps of Engineers.
Army Corps of Engineers. (1963b). *Technical report on barriers: A part of the comprehensive survey of San Francisco Bay and tributaries, California.* Appendix H, volume 1: *Hydraulic model studies.* San Francisco: Army Corps of Engineers.
Army Corps of Engineers. (1981). *San Francisco Bay – delta tidal hydraulic model: User's manual.* San Francisco: Army Corps of Engineers.
Attneave, F. (1950). Dimensions of similarity. *American Journal of Psychology, 63,* 516–556.
Barnard, P. (2011). Anthropogenic influences on shoreline and nearshore evolution in the San Francisco Bay coastal system. *Estuarine, Coastal and Shelf Science, 92*(1), 195–204.
Batterman, R. W. (2001). *The devil in the details: Asymptotic reasoning in explanation, reduction, and emergence.* New York: Oxford University Press.
Batterman, R. W. (2002). Asymptotics and the role of minimal models. *British Journal for the Philosophy of Science, 53,* 21–38.
Bender, C. M., & Orszag, S. A. (1999). *Advanced mathematical methods for scientists and engineers.* New York: Springer.
Berlekamp, E. R., Conway, J. H., & Guy, R. K. (1982). What is life? In *Winning ways for your mathematical plays* (chap. 25). New York: Academic Press.
Berryman, A. A. (1992). The origins and evolution of predator–prey theory. *Ecology, 73*(5), 1530–1535.
Borges, J. L. (1998). On exactitude in science. In *Collected fictions.* Translated by Andrew Hurley. Harmondsworth: Penguin.
Black, M. (1962). *Models and metaphors.* Ithaca: Cornell University Press.
Briggs, C. J., & Hoopes, M. F. (2004). Stabilizing effects in spatial parasitoid–host and predator–prey models: A review. *Theoretical Population Biology, 65,* 299–315.
Bueno, O., French, S., & Ladyman, J. (2002). On representing the relationship between the mathematical and the empirical. *Philosophy of Science, 69,* 497–518.
Bull, J. J., & Pease, C. M. (1989). Combinatorics and variety of mating-type systems. *Evolution, 43*(3), 667–671.
Campbell, N. R. (1957). *Foundations of science.* New York: Dover.
Cartwright, N. (1983). *How the laws of physics lie.* Oxford: Oxford University Press.
Cartwright, N. (1989). *Nature's capacities and their measurement.* Oxford: Oxford University Press.
Casselton, L. A. (2002). Mate recognition in fungi. *Heredity, 88*(2), 142–147.
Collins, J. D., Hall, E. J., & Paul, L. A. (2004). *Causation and counterfactuals.* Cambridge, MA: MIT Press.
Contessa, G. (2010). Scientific models and fictional objects. *Synthese, 172*(2), 215–219.

Crow, J. (1992). An advantage of sexual reproduction in a rapidly changing environment. *Journal of Heredity, 83*(3), 169.

da Costa, N. C. A., & French, S. (2003). *Science and partial truth.* Oxford: Oxford University Press.

Davis, W. (2010). Implicature. In E. N. Zalta (Ed.), *The Stanford encyclopedia of philosophy* (Winter 2010). http://plato.stanford.edu/archives/win2010/entries/implicature/.

Dennett, D. C. (1991). Real patterns. *Journal of Philosophy, 88*(1), 27–51.

Diamond, J. (1999). *Guns, germs, and steel: The fate of human societies.* New York: Norton.

Downes, S. M. (1992). The importance of models in theorizing: A deflationary semantic view. *PSA: Proceedings of the Biennial Meeting of the Philosophy of Science Association, 1,* 142–153.

Eco, U. (1991). Small worlds. In *The limits of interpretation.* Indianapolis: Indiana University Press.

Eddington, A. S. (1927). *The nature of the physical world.* Cambridge: Cambridge University Press.

Eigen, M., & Winkler, R. (1983). *Laws of the game: How the principles of nature govern chance.* New York: Harper & Row.

Elliott-Graves, A. (2012). Abstract and complete. PhilSci Archive. Retrieved from http://philsci-archive.pitt.edu/id/eprint/9274/.

Elton, C., & Nicholson, M. (1942). The ten-year cycle in numbers of the lynx in Canada. *Journal of Animal Ecology, 11*(2), 215–244.

Ermentrout, G., & Edelstein-Keshet, L. (1993). Cellular automata approaches to biological modeling. *Journal of Theoretical Biology, 160*(1), 97–133.

Ewens, W. J. (1963). The mean time for absorption in a process of genetic type. *Journal of the Australian Mathematical Society, 3,* 375–383.

Feynman, R. P., Leighton, R. B., & Sands, M. L. (1989). *The Feynman lectures on physics.* Redwood City, CA: Addison-Wesley.

Fisher, R. A. (1930). *The genetical theory of natural selection.* Oxford: The Clarendon Press.

Forber, P. (2010). Confirmation and explaining how possible. *Studies in History and Philosophy of Science Part C, 41*(1), 32–40.

Foresman, J. B., & Frisch, A. (1996). *Exploring chemistry with electronic structure methods* (2nd edn.). Pittsburgh: Gaussian.

Forseth, Jr., I. N., & Innis, A. F. (2004). Kudzu (*Pueraria montana*): History, physiology, and ecology combine to make a major ecosystem threat. *Critical Reviews in Plant Sciences, 23*(5), 401–413.

French, S. (2006). Structure as a weapon of the realist. *Proceedings of the Aristotelian Society, 106,* 1–19.

French, S. (2010). Keeping quiet on the ontology of models. *Synthese, 172*(2), 231–249.

French, S., & Ladyman, J. (1998). Semantic perspective on idealization in quantum mechanics. In N. Shanks (Ed.), *Idealization in contemporary physics.* Poznan studies in the philosophy of the sciences and the humanities (Vol. 63, pp. 51–73). Amsterdam: Rodopi.

French, S., & Ladyman, J. (2003). Remodelling structural realism: Quantum physics and the metaphysics of structure. *Synthese, 136,* 31–56.

Friedman, M. (1974). Explanation and scientific understanding. *Journal of Philosophy, 71,* 5–19.

Friend, S. (2009). Unpublished comments at Models and Fictions Conference, University of London Institute of Philosophy.
Frigg, R. (2010). Models and fiction. *Synthese, 172*(2), 251–268.
Gao, J., Liu, H., & Kool, E. T. (2005). Assembly of the complete eight-base artificial genetic helix, xDNA, and its interaction with the natural genetic system. *Angewandte Chemical International Edition English, 44*(20), 3118–3122.
Gardner, M. (1970). The fantastic combinations of John Conway's new solitaire game "Life". *Scientific American, 223*, 120–123.
Gardner, M. (1983). The Game of Life, parts I–III. In *Wheels, life, and other mathematical amusements* (chap. 20–22). New York: Freeman.
Gentner, D., & Markman, A. (1998). Structure mapping in analogy and similarity. *Mind Readings: Introductory Selections on Cognitive Science.*
Gentner, D., & Markman, A. (1994). Structural alignment in comparison: No difference without similarity. *Psychological Science, 5*, 152–158.
Giere, R. N. (1988). *Explaining science: A cognitive approach.* Chicago: University of Chicago Press.
Giere, R. N. (2009). Is computer simulation changing the face of experimentation? *Philosophical Studies, 143*(1), 59–62.
Gilpin, M. E. (1973). Do hares eat lynx? *The American Naturalist, 107*(957), 727–730.
Gleitman, L. R., Gleitman, H., Miller, C., & Ostrin, R. (1996). Similar, and similar concepts. *Cognition, 58*, 321–376.
Godfrey-Smith, P. (2006). The strategy of model-based science. *Biology and Philosophy, 21*, 725–740.
Godfrey-Smith, P. (2009). Models and fictions in science. *Philosophical Studies, 143*(1), 101–116.
Godfrey-Smith, P., & Lewontin, R. (1993). The dimensions of selection. *Philosophy of Science, 60*(3), 373–395.
Goldstone, R. L. (1994). The role of similarity in categorization: Providing a groundwork. *Cognition, 52*, 125–157.
Goldstone, R. L., Medin, D., & Gentner, D. (1991). Relational similarity and the non-independence of features in similarity judgments. *Cognitive Psychology, 23*, 222–262.
Goodman, N. (1972). Seven strictures on similarity. In *Problems and projects.* Indianapolis: Bobbs-Merril.
Goodwin, R. (1967). A growth cycle. In C. H. Feinstein (Ed.), *Socialism, capitalism, and economic growth.* Cambridge: Cambridge University Press.
Grice, H. P. (1975). Logic and conversation. In P. Cole & J. Morgan (Eds.), *Syntax and semantics, 3: Speech acts* (pp. 41–58). New York: Academic Press.
Grice, H. P. (1981). Presupposition and conversational implicature. In P. Cole (Ed.), *Radical pragmatics* (pp. 183–98). New York: Academic Press.
Griesemer, J. R. (2003). Three-dimensional models in philosophical perspective. In S. de Chadarevian & N. Hopwood (Eds.), *Displaying the third dimension: Models in the sciences, technology and medicine.* Stanford: Stanford University Press.
Griesemer, J. R., & Wade, M. J. (1988). Laboratory models, causal explanation, and group selection. *Biology and Philosophy, 3*, 67–96.
Grimm, V., & Railsback, S. F. (2005). *Individual-based modeling and ecology.* Princeton: Princeton University Press.
Grüne-Yanoff, T. (2011). Evolutionary game theory, interpersonal comparisons and natural selection: A dilemma. *Biology and Philosophy, 26*(5), 637–654.

Guala, F. (2002). Models, simulations, and experiments. In L. Magnani & N. Nersessian (Eds.), *Model-based reasoning: Science, technology, values* (p. 59–74). New York: Kluwer.

Hahn, U., Chater, N., & Richardson, L. (2003). Similarity as transformation. *Cognition, 87*, 1–32.

Hahn, U., & Ramscar, M. (2001). *Similarity and categorization.* Oxford: Oxford University Press.

Haldane, J. B. S. (1932). *The causes of evolution.* London: Longmans, Green.

Hanski, I., Henttonen, H., Korpimaki, E., Oksanen, L., & Turchin, P. (2001). Small rodent dynamics and predation. *Ecology, 82*(6), 1505–1520.

Harder, J. A. (1957). *An electric analog model study of tides in the delta region of California.* Berkeley: University of California, Institute of Engineering Research.

Hartmann, S. (1998). Idealization in quantum field theory. In N. Shanks (Ed.), *Idealization in contemporary physics.* (pp. 99–122). Amsterdam: Rodopi.

Hempel, C. G. (1965). Aspects of scientific explanation. In *Aspects of scientific explanation and other essays* (pp. 331–496). New York: The Free Press.

Hendry, R., & Psillos, S. (2007). How to do things with theories: an interactive view of language and models in science. In J. Brzeziński, A. Klawiter, T. Kuipers, K. Łastowski, K. Paprzycka, & P. Przybysz (Eds.), *The courage of doing philosophy: Essays dedicated to Leszek Nowak* (pp. 59–115). Amsterdam: Rodopi.

Heppner, F., & Grenander, U. (1990). A stochastic nonlinear model for coordinated bird flocks. In S. Krasner (Ed.), *The ubiquity of chaos.* Washington, DC: AAAS Publications.

Hesse, M. B. (1966). *Models and analogies in science.* South Bend, IN: University of Notre Dame Press.

Hoffmann, R., Minkin, V. I., & Carpenter, B. K. (1996). Ockham's razor and chemistry. *Bulletin de la Société Chimique de France, 133*, 117–130.

Holling, C. S. (1959). The components of predation as revealed by a study of small mammal predation of the European pine sawfly. *Canadian Journal of Entomology, 91*, 293–320.

Huggins, E. M., & Schultz, E. A. (1967). San Francisco Bay in a warehouse. *Journal of the IEST, 10*(5), 9–16.

Huggins, E. M., & Schultz, E. A. (1973). The San Francisco Bay and Delta model. *California Engineer, 51*(3), 11–23.

Hughes, R. I. G. (1997). Models and representation. *Philosophy of Science, 64*, S325–S336.

Humphreys, P. (2007). *Extending ourselves: Computational science, empiricism, and scientific method.* New York: Oxford University Press.

Hurst, L. D. (1996). Why are there only two sexes? *Proceedings of the Royal Society of London B: Biological Sciences, 263*(1369), 415–422.

Hurst, L. D., & Peck, J. R. (1996). Recent advances in understanding of the evolution and maintenance of sex. *Trends in Ecology & Evolution, 11*(2), 46–52.

Ilachinski, A. (2001). *Cellular automata: A discrete universe.* Singapore: World Scientific.

Ising, E. (1925). Beitrag zur theorie des ferromagnetismus. *Zeitschrift für Physik A: Hadrons and Nuclei, 31*, 253–258.

Jackson, W. T., & Paterson, A. M. (1977). *The Sacramento–San Joaquin Delta: The evolution and implementation of water policy.* Davis: California Water Resources Center, University of California.

Jones, M. R. (2005). Idealization and abstraction: A framework. In M. Jones & N. Cartwright (Eds.), *Idealization XII: Correcting the model. Idealization and abstraction in the sciences* (pp. 173–217). Amsterdam: Rodopi.

Jurrell, D. J. (2005). Local spatial structure and predator–prey dynamics: Counterintuitive effects of prey enrichment. *American Naturalist, 166*, 354–367.

Justus, J. (2006). Loop analysis and qualitative modeling: limitations and merits. *Biology & Philosophy, 21*, 647–666.

Kant, I. (1998). *Critique of pure reason* (P. Guyer & A. Wood, Eds.). Cambridge, UK: Cambridge University Press.

Karlin, S., & Feldman, M. (1969). Linkage and selection: New equilibrium properties of the two-locus symmetric viability model. *Proceedings of the National Academy of Sciences of the United States of America, 62*(1), 70.

Kauffman, S. A. (1993). *The origins of order.* Oxford: Oxford University Press.

Kemp, C., Bernstein, A., & Tenenbaum, J. B. (2005). A generative theory of similarity. In *Proceedings of the 27th annual conference of the Cognitive Science Society.*

Kimbrough, S. O. (2003). Computational modeling and explanation: Opportunities for the information and management sciences. In H. K. Bhargava & N. Ye (Eds.), *Computational modeling and problem solving in the networked world: Interfaces in computing and optimization* (pp. 31–57). Boston, MA: Kluwer.

Kimura, M., & Ohta, T. (1969). The average number of generations until fixation of a mutant gene in a finite population. *Genetics, 61*(3), 763–771.

Kitcher, P. (1981). Explanatory unification. *Philosophy of Science, 48*, 507–531.

Kitcher, P. (1993). *The advancement of science.* Oxford: Oxford University Press.

Kittel, C. (1980). *Thermal physics.* New York: W. H. Freeman.

Kline, S. J. (1986). *Similitude and approximation theory.* Berlin: Springer-Verlag.

Knuuttila, T. (2009a). Isolating representations versus credible constructions? Economic modelling in theory and practice. *Erkenntnis, 70*, 59–80.

Knuuttila, T. (2009b). Representation, idealization, and fiction in economics. In M. Suárez (Ed.), *Fictions in science* (pp. 205–231). London: Routledge.

Ladyman, J. (1998). What is structural realism? *Studies in History and Philosophy of Science, 29*, 409–424.

Langton, C. G. (1995). *Artificial life: An overview.* Cambridge, MA: MIT Press.

Levine, I. N. (2002). *Physical chemistry* (5th ed.). Boston: McGraw-Hill.

Levins, R. (1962). Theory of fitness in a heterogeneous environment, I. The fitness set and adaptive function. *American Naturalist, 96*(861), 361–373.

Levins, R. (1966). The strategy of model building in population biology. In E. Sober (Ed.), *Conceptual issues in evolutionary biology* (1st ed., pp. 18–27). Cambridge, MA: MIT Press.

Levins, R., & MacArthur, R. H. (1966). The maintenance of genetic polymorphism in a spatially heterogenous environment. *American Naturalist, 100*, 585–589.

Levy, A. (2012). *Fictional models de novo and de re.* PhilSci Archive. Retrieved from http://philsci-archive.pitt.edu/id/eprint/9075/.

Lewis, D. (1978). Truth in fiction. *American Philosophical Quarterly, 15*, 37–46.

Lewis, G. N. (1916). The atom and the molecule. *Journal of the American Chemical Society, 38*(4), 762–785.

Lewontin, R. C. (1974). *The genetic basis of evolutionary change.* New York: Columbia University Press.

Liu, H., Gao, J., Lynch, S. R., Saito, Y. D., Maynard, L., & Kool, E. T. (2003). A four-base paired genetic helix with expanded size. *Science*, *302*(5646), 868–871.
Lloyd, E. A. (1984). A semantic approach to the structure of population genetics. *Philosophy of Science*, *51*(2), 242–264.
Lloyd, E. A. (1994). *The structure and confirmation of evolutionary theory* (2nd ed.). Princeton: Princeton University Press.
Lotka, A. J. (1956). *Elements of mathematical biology*. New York: Dover.
Lustick, I. (2011). Secession of the center: A virtual probe of the prospects for Punjabi secessionism in Pakistan and the secession of Punjabistan. *Journal of Artificial Societies and Social Simulation*, *14*(1), 7.
Matthewson, J. (2012). *Generality and the limits of model-based science*. Unpublished doctoral dissertation, Australian National University.
Matthewson, J., & Weisberg, M. (2009). The structure of tradeoffs in model building. *Synthese*, *170*(1), 169–190.
May, R. M. (1972). Limit cycles in predator–prey communities. *Science*, *177*(4052), 900–902.
May, R. M. (2001). *Stability and complexity in model ecosystems*. Princeton: Princeton University Press.
May, R. M. (2004). Uses and abuses of mathematics in biology. *Science*, *303*, 790–793.
Maynard Smith, J. (1974). *Models in ecology*. Cambridge: Cambridge University Press.
Maynard Smith, J. (1989). *Evolutionary genetics*. Oxford: Oxford University Press.
McMullin, E. (1985). Galilean idealization. *Studies in the History and Philosophy of Science*, *16*, 247–273.
Miller, J. H., & Page, S. E. (2007). *Complex adaptive systems: An introduction to computational models of social life*. Princeton: Princeton University Press.
Mitchell, S. D. (2000). Dimensions of scientific law. *Philosophy of Science*, *67*(2), 242–265.
Moran, P. (1953). The statistical analysis of the Canadian lynx cycle. I. Structure and prediction. *Australian Journal of Zoology*, *1*, 163–173.
Morgan, M. S., & Morrison, M. (1999). Models as mediating instruments. In M. S. Morgan & M. Morrison (Eds.), *Models as mediators* (pp. 10–37). Cambridge: Cambridge University Press.
Morrison, M. (2009). Models, measurement and computer simulation: The changing face of experimentation. *Philosophical Studies*, *143*(1), 33–57.
Muldoon, R., Smith, T., & Weisberg, M. (2012). Segregation that no one seeks. *Philosophy of Science*, *79*, 38–62.
Nayfeh, A. H. (2000). *Perturbation methods*. New York: John Wiley.
Nowak, L. (1972). Laws of science, theories, measurement. *Philosophy of Science*, *34*, 533–548.
Nowak, M. A. (2006). *Evolutionary dynamics: Exploring the equations of life*. Cambridge, MA: Harvard University Press.
Odenbaugh, J. (2003). Complex systems, trade-offs and mathematical modeling: Richard Levins' "Strategy of Model Building in Population Biology" revisited. *Philosophy of Science*, *70*, 1496–1507.
Odenbaugh, J. (in press). True lies: Robustness and idealizations in ecological explanations. *Philosophy of Science*.

Orzack, S. H. (2005). What, if anything, is "The Strategy of Model Building in Population Biology"? A comment on Levins (1966) and Odenbaugh (2003). *Philosophy of Science, 72*, 479–485.

Orzack, S. H., & Sober, E. (1993). A critical assessment of Levins's "The Strategy of Model Building in Population Biology (1966)". *Quarterly Review of Biology, 68*(4), 533–546.

Parke, E. (in press). What could arsenic bacteria teach us about life? *Biology and Philosophy.*

Parker, J. D. (2004). A major evolutionary transition to more than two sexes? *Trends in Ecology & Evolution, 19*(2), 83–86.

Parker, W. S. (2009). Does matter really matter? Computer simulations, experiments, and materiality. *Synthese, 169*(3), 483–496.

Pincock, C. (2005). Overextending partial structures: Idealization and abstraction. *Philosophy of Science, 72*(5), 1248–1259.

Pincock, C. (2011). *Mathematics and scientific representation.* New York: Oxford University Press.

Poundstone, W., & Wainwright, R. T. (1985). *The recursive universe: Cosmic complexity and the limits of scientific knowledge* (1st ed.). New York: Morrow.

Puccia, C. J., & Levins, R. (1985). *Qualitative modeling of complex systems: An introduction to loop analysis and time averaging.* Cambridge, MA: Harvard University Press.

Quine, W. (1969). Natural kinds. In *Ontological relativity and other essays.* New York: Columbia University Press.

Raper, J. R. (1966). *Genetics of sexuality in higher fungi.* New York: Ronald Press.

Reber, J. (1959). "Our perpetual gift to California and the nation: A master plan for the vast San Francisco region" (National Archives and Records Administration. Pacific Region, Papers of John Reber, Location 2126E-G, Accn. 77-94-09).

Rendell, P. (2002). Turing universality in the game of life. In A. Adamatzky (Ed.), *Collision-based computing* (pp. 513–539). London: Springer.

Reynolds, C. W. (1987). Flocks, herds, and schools: A distributed behavioral model, in computer graphics. *SIGGRAPH '87 Conference Proceedings, 21*(4), 25–34.

Rice, C., & Smart, J. (2011). Interdisciplinary modeling: A case study of evolutionary economics. *Biology and Philosophy, 26*(5), 655–675.

Richardson, R. C. (2006). Chance and the patterns of drift: A natural experiment. *Philosophy of Science, 73*, 642–654.

Ricklefs, R. E., & Miller, G. L. (2000). *Ecology* (4th ed.). New York: Freeman.

Rothman, D. H., & Zaleski, S. (2004). *Lattice-gas cellular automata: simple models of complex hydrodynamics* (1st pbk. ed., Vol. 5). Cambridge, UK: Cambridge University Press.

Roughgarden, J. (1979). *Theory of population genetics and evolutionary ecology: An introduction.* New York: Macmillan.

Roughgarden, J. (1997). *Primer of ecological theory.* Upper Saddle River, NJ: Prentice Hall.

Royama, T. (1971). A comparative study of models for predation and parasitism. *Researches on Population Ecology, Sup. #1*, 1–91.

Ryan, M. L. (1980). Fiction, non-factuals, and the principle of minimal depature. *Poetics, 9,* 403–422.

Scanlan, J., Berman, D., & Grant, W. (2006). Population dynamics of the European rabbit (*Oryctolagus cuniculus*) in north eastern Australia: Simulated responses to control. *Ecological modelling, 196*(1–2), 221–236.

Schelling, T. C. (1978). *Micromotives and macrobehavior.* New York: Norton.

Schulz, L., Gopnik, A., & Glymour, C. (2007). Preschool children learn about causal structure from conditional interventions. *Developmental Science, 10*(3), 322–332.

Seger, J., & Brockmann, H. J. (1987). What is bet-hedging? *Oxford Surveys in Evolutionary Biology, 4*, 181–211.

Shepard, R. N. (1980). Multidimensional scaling, tree-fitting, and clustering. *Science, 210*, 390–398.

Shepard, R. N. (1987). Toward a universal law of generalization for psychological science. *Science, 237*, 1317–1323.

Shepard, R. N., & Metzler, J. (1971). Mental rotation of three-dimensional objects. *Science, 171*(3972), 701–703.

Singer, D. J. (2008). *The problem of scientific modeling for structural realism and the semantic view of theories.* Unpublished honors thesis, University of Pennsylvania.

Sneed, J. (1971). *The logical structure of mathematical physics.* Dordrecht, Holland: Reidel.

Stalnaker, R. (2002). Common ground. *Linguistics and Philosophy, 25*, 701–721.

Stenseth, N. C., Falck, W., Bjørnstad, O. N., & Krebs, C. J. (1997). Population regulation in snowshoe hare and Canadian lynx: Asymmetric food web configurations between hare and lynx. *Proceedings of the National Academy of Sciences, 94*(10), 5147–5152.

Sterrett, S. G. (2005). *Wittgenstein flies a kite: A story of models of wings and models of the world.* New York: Pi Press.

Strevens, M. (1998). Inferring probabilities from symmetries. *Noûs, 32*(2), 231–246.

Strevens, M. (2004). The causal and unification approachs to explanation unified—causally. *Noûs, 38*, 154–179.

Strevens, M. (2008). *Depth: An account of scientific explanation.* Cambridge, MA: Harvard University Press.

Suárez, M. (2004). An inferential conception of scientific representation. *Philosophy of Science, 71*(5), 767–779.

Suárez, M. (2009). *Fictions in science: Philosophical essays on modeling and idealization* (Vol. 4). New York: Routledge.

Sugden, R. (2002). Credible worlds: The status of the theoretical models in economics. In U. Maki (Ed.), *Fact and fiction in economics: Models, realism, and social construction* (pp. 107–136). Cambridge, UK: Cambridge University Press.

Suppe, F. (1977a). The search for philosophic understanding of scientific theories. In F. Suppe (Ed.), *The structure of scientific theories* (2nd ed.). Chicago: University of Illinois Press.

Suppe, F. (Ed.). (1977b). *The structure of scientific theories.* Chicago: University of Illinois Press.

Suppe, F. (1989). *The semantic conception of theories and scientific realism.* Chicago: University of Illinois Press.

Suppes, P. (1960a). A comparison of the meaning and use of models in mathematics and the empirical sciences. *Synthese, 12*, 287–300.

Suppes, P. (1960b). Models of data. In E. Nagel & P. Suppes (Eds.), *Logic, methodology and the philosophy of science: Proceedings of the 1960 international congress* (pp. 251–261). Stanford: Stanford University Press.

Tanenbaum, A. S., & Van Steen, M. (2002). *Distributed systems: Principles and paradigms.* Upper Saddle River, NJ: Prentice Hall.

Teller, P. (2001). Twilight of the perfect model model. *Erkenntnis, 55*(3), 393–415.

Thomasson, A. L. (1999). *Fiction and metaphysics.* Cambridge, UK: Cambridge University Press.

Thomson-Jones, M. (1997). *Models and the semantic view*. PhilSci Archive. Retrieved from http://philsci-archive.pitt.edu/8994/

Thomson-Jones, M. (2006). Models and the semantic view. *Philosophy of Science, 73*(5), 524–535.

Toon, A. (2010). The ontology of theoretical modelling: Models as make-believe. *Synthese, 172*(2), 301–315.

Toon, A. (2012). *Models as make-believe: Imagination, fiction and scientific representation*. Basingstoke, UK: Palgrave Macmillan.

Tversky, A. (1977). Features of similarity. *Psychological Review, 84*, 327–352.

Tversky, A., & Gati, I. (1978). Studies of similarity. In E. Rosch & B. Lloyd (Eds.), *Cognition and categorization*. Hillsdale, NJ: Erlbaum.

Vaihinger, H. (1911). *The philosophy of "as if"*. London: Kegan Paul.

van Fraassen, B. C. (1980). *The scientific image*. Oxford: Oxford University Press.

Vichniac, G. Y. (1984). Simulating physics with cellular automata. *Physica D: Nonlinear Phenomena, 10*(1–2), 96–116.

Volterra, V. (1926a). Fluctuations in the abundance of a species considered mathematically. *Nature, 118*, (558–560.).

Volterra, V. (1926b). Variazioni e fluttuazioni del numero d'individui in specie animali conviventi. *Memorie Della R. Accademia Nazionale Dei Lincei, II*, 5–112.

Walton, K. L. (1990). *Mimesis as make-believe: On the foundations of the representational arts*. Cambridge, MA: Harvard University Press.

Watts, D. J., & Strogatz, S. H. (1998). Collective dynamics of "small-world" networks. *Nature, 393*, 440–442.

Weber, M. (2005). *Philosophy of experimental biology*. Cambridge, UK: Cambridge University Press.

Weisberg, D. S. (2008). *The creation and comprehension of fictional worlds*. Unpublished doctoral dissertation, Yale University.

Weisberg, D. S., & Goodstein, J. (2009). What belongs in a fictional world? *Journal of Cognition and Culture, 9*, 69–78.

Weisberg, M. (2003). *When less is more: Tradeoffs and idealization in model building*. Unpublished doctoral dissertation, Stanford University.

Weisberg, M. (2006). Forty years of "The Strategy": Levins on model building and idealization. *Biology and Philosophy, 21*(5), 623–645.

Weisberg, M. (2007a). Three kinds of idealization. *Journal of Philosophy, 104*(12), 639–659.

Weisberg, M. (2007b). Who is a modeler? *British Journal for the Philosophy of Science, 58*, 207-233.

Weisberg, M., & Reisman, K. (2008). The robust Volterra principle. *Philosophy of Science, 75*, 106–131.

Westheimer, F. H. (1987). Why nature chose phosphates. *Science, 235*(4793), 1173–1178.

Wimsatt, W. C. (1980). Randomness and perceived randomness in evolutionary biology. *Synthese, 43*, 287–329.

Wimsatt, W. C. (1981). Robustness, reliability, and overdetermination. In M. Brewer & B. Collins (Eds.), *Scientific inquiry and the social sciences* (pp. 124–163). San Francisco: Jossey-Bass.

Wimsatt, W. C. (1987). False models as means to truer theories. In M. Nitecki & A. Hoffmann (Eds.), *Neutral models in biology* (pp. 23–55). Oxford: Oxford University Press.

Wimsatt, W. C. (2007). *Re-engineering philosophy for limited beings*. Cambridge, MA: Harvard University Press.
Winsberg, E. B. (2010). *Science in the age of computer simulation*. Chicago: University of Chicago Press.
Winther, R. G. (2006). Parts and theories in compositional biology. *Biology & Philosophy, 21*, 471–499.
Wisdom, J., & Holman, M. (1991). Symplectic maps for the n-body problem. *Astronomical Journal, 102*, 1528–1538.
Wolfram, S. (1983). Statistical mechanics of cellular automata. *Reviews of Modern Physics, 55*(3), 601–644.
Woodward, J. (2003). *Making things happen*. Oxford: Oxford University Press.

■ INDEX

Abstract direct representation, 64, 74, 130
Abstraction, 18–19, 23, 34, 91–92, 94, 102, 116, 172
   of targets. *See* Target systems, generalized
Agent-based models. *See* Models, computational, agent-based
Alcohol, molecular structure of, 142
Algorithm. *See* Structures, computational, algorithms
Army Corps of Engineers, 1–3, 4, 7–9, 25, 34, 38–39, 84–88, 162–163, 171
Attneave, Frank, 143

Batterman, Robert, 101, 102
Best model ever. *See* San Francisco Bay model
Bohr model, 21, 22
Boids model, 120
Boyle's law, 101. *See also* Ideal gas model

Carrying capacity, 10, 125
Cartwright, Nancy, 102, 142, 143, 155
Causal information in models, 72
Cellular automata, 129–131
Ceteris paribus clauses, 158–159, 167–168
Chemical bonding models, 107
Construal, 39–42, 76–77, 149, 171
  assignment, 39–40, 171
  fidelity criteria, 39, 41–42, 73, 93, 125, 172
    dynamical, 41
    representational, 41
  scope, 39, 40–41, 91, 149, 172
  setting, 117
Conway, J. H., 129
Crick, Francis, 24, 38
Crow, J. F., 115–117

D'Ancona, Umberto, 3
Dennett, Daniel, 131
Desiderata for model-world relations, 135
  adjudication, 137, 155
  context, 137, 141
  idealization, 136, 139, 155
  maximality, 135, 147, 154
  qualitative, 136, 141, 154
  richness, 136, 154
  scalar, 136, 141, 154
  tractability, 137, 155
Desiderata of models, 103–104. *See also* Representational ideals
Diamond, Jared, 25
Discretized model equations, 83
DNA model, 24, 38
Downes, Steve, 18
Drosophila melanogaster, 16, 24

Ecology, 79, 92, 103, 111
Eukaryotic cell, model, 18, 19
Experiments, 17, 25, 82
  natural experiments, 25
Exponential growth model, 80, 89, 124–125

Feldman, Marcus, 65
Feynman's Ratchet, 126–127
Fictions, 46
  as possibilities, 51–52
  problem of accounting for the practice of modeling, 63–64
  problem of representational capacity, 61–63
  problem of variation for, 56–61, 64–67
  as products of imaginations, 51–52, 53–54
  simple fictions account of models, 50–51, 52, 53, 54, 55
  Waltonian fictionalism, 53, 55, 67

Fisher, R. A., 121, 131–132
Fitness function, 88
Focal properties of stories, 57–58, 59, 60
Folk ontology of models, 68–70, 73
Frigg, Roman, 46, 53–56, 58, 61

Galileo. *See* Idealization, Galilean
Game of Life, 129–131
Giere, Ronald, 33–34, 142, 143, 155
Godfrey-Smith, Peter, 46, 48, 49, 50, 51, 52, 53, 56, 67, 142
Goodman, Nelson, 142–143
Goodwin, Richard, 77–78

Haldane, J. B. S., 115
Harmonic oscillator model, 39, 47, 138–139
Hartmann, Stephan, 101–102
Harvester ants, 133
Hoffman, Roald, 108
How-possibly questions. *See* Modeling, how-possibly
Humphreys, Paul, 81–82
Hurst, Laurence, 132–133
Hurst, Lester, 92, 124

Ideal gas model, 140–141. *See also* Boyle's law
Idealization, 5, 18–19, 55, 97, 98, 113, 156, 174. *See also* Desiderata for model-world relations
  discretization, 83
  distortion, 98, 99
  Galilean, 99–100, 102, 103, 110–111, 112, 174
  minimalist, 100–103, 104, 111, 112, 174
  multiple-models, 103–105, 111–112, 113, 174
  parameterization, 83
  representational ideals of. *See* Representational ideals
Individual-based models. *See* Models, computational, agent-based

Initial conditions. *See* State, initial
Interpretation of models. *See* Construal
Ising, Ernst. *See* Ising model
Ising model, 100

Kairetic account of scientific explanation, 101
  causal entailment, 101
  difference-makers, 101
Karlin, Samuel, 65
Kitcher, Philip, 154
Kool, Eric, 122–124
Kudzu plant, 16

Levins, Richard, 103–104, 112, 156–158, 167
Levy, Arnon, 46, 55–56, 64
Lewis, David, 52
Linkage disequilibrium, 65
Logistic growth model, 125, 169
Lotka, Alfred, 10. *See also* Lotka-Volterra model of predation
Lotka-Volterra model of predation, 3–4, 10–13, 14–15, 18, 27–28, 30, 37–38, 39, 44, 49–50, 58–60, 61–62, 63, 68, 75, 80, 81, 89, 93, 125, 136, 149, 161
  equilibrium values, 12–13, 81
  functional response, 11
  individual-based, 38, 62, 163–166
  numerical response, 11
  spatial structure in, 60
  Volterra principle, 12, 84, 89, 161, 166
  Volterra property, 12, 13, 160, 161, 166
  use in economics, 77–78
Lustick, Ian, 131

May, Robert, 119
Maynard Smith, John, 48–49, 68, 125
McMullin, Ernan, 99
Mental rotation, 17. *See also* Models, verbal
Metzler, Jacqueline, 17–18
Middle Earth, 53, 57

Model descriptions, 31–39, 172
　computational, 38
　concrete, 35–36
　instantiations of, 37
　mathematical, 37, 57–58
　relation to models, 34–35
Model-based theorizing. *See* Modeling
Modeling
　face-value practice of, 48–49, 64, 73
　generalized, 64, 114–121
　how-possibly, 118–119, 151, 173
　hyperaccurate, 150–151
　hypothetical, 114, 121–129
　interpretation in. *See* Construal
　mechanistic, 151–152
　minimal. *See* Models, minimal
　practice of, 4, 16, 19–20, 63–64, 154
　target-directed. *See* Target-directed modeling
　targetless, 64, 114, 129–131
Models, 171, 174–175
　analytical solutions to, 80, 82
　computational, 7, 13–14, 15, 19, 20, 30, 31, 43, 44
　　agent-based, 13–14, 70, 120, 163–164
　concrete, 7–9, 15, 16, 18–19, 20–22, 24–25, 33, 43, 71
　construals of. *See* Construal
　dynamical, 15, 26, 28–29
　grids of, 83
　idealized exemplars, 16, 18–19
　interpreted structure in, 15
　Lotka-Volterra. *See* Lotka-Volterra model of predation
　mathematical, 7, 10–13, 15, 18–19, 20–22, 25–29, 30, 43, 44, 69, 70–73, 95
　　dynamical, 15, 26, 28–29, 95
　as mathematical objects, 47
　minimal, 100, 107, 119–120, 151. *See also* Idealization, minimalist
　model organisms, 16–17, 24
　narrative. *See* Models, verbal
　numerical analysis of, 81, 82
　propositional, 21
　representational capacity of. *See* Representational capacity
　resolution of, 83
　San Francisco Bay. *See* San Francisco Bay model
　Schelling's segregation model. *See* Schelling's segregation model
　structures in. *See* Structures
　verbal, 16, 17–18, 19
Model–world relations, 23, 56, 96, 118, 135, 149–150, 155, 173, 175. *See also* Target-directed modeling, model-target comparison
　desiderata for. *See* Desiderata for model-world relations
　empirical adequacy, 47
　homomorphism, 137, 138
　isomorphism, 47, 137–140, 150
　partial isomorphism, 137, 140–142
　similarity. *See* Similarity
Monte Carlo methods, 82
Moore neighborhood, 13, 14
Morgan, Mary, 46n1
Morrison, Margaret, 46n1
Mutual belief principle, 59–60. *See also* Fictions

Nowak, Martin, 55

Orzack, Steven, 37, 157–158, 167, 169

Perpetual motion, 126–128
Perturbation theory, 89
Phase space. *See* State space
Pincock, Chris, 20–21
Potential energy surfaces, 28
Predation, 10, 92, 111. *See also* Lotka-Volterra model of predation

Quine, W. V. O., 142

Reality principle, 58. *See also* Fictions
Reber, John. *See* Reber Plan
Reber Plan, 1, 7–9, 17, 25, 38, 84–88, 171

Reisman, Ken, 30, 162
Representational capacity, 43, 75, 171
  dynamical sufficiency, 43
  mechanistic adequacy, 44
Representational ideals, 105, 112, 174
  1-causal, 107–108, 111, 120
  completeness, 105–106, 109, 111, 150
    fidelity rules, 106
    inclusion rules, 106
  generality, 109–110
    a-generality, 109
    p-generality, 109
  maxout, 109
  simplicity, 107, 108, 111
RNA replication, 48–49
Robustness. *See* Robustness analysis
Robustness analysis, 5, 156–160, 174
  parameter, 160–161
  representational, 160, 162–166
  robust theorems, 157–159, 174
  role in confirmation theory, 167–170
    low-level confirmation, 169
  structural, 160, 161–162
Roughgarden, Joan, 115, 119

San Francisco Bay model, 1–3, 7–9, 14–15, 16, 17, 21, 22, 24, 25, 31–34, 35–36, 38–39, 42, 80, 82, 84–88, 94, 136, 147–148, 149, 153, 162–163, 167–168, 171
Schelling, Thomas, 13–14
Schelling's segregation model, 13–14, 20, 23, 30, 31, 88, 94, 118–119, 136, 139–140, 141, 145–146, 148, 158, 168, 169
Scientific explanation, 101, 108, 125–126
Segregation. *See* Schelling's segregation model
Semantic view of theories, 26
  set-theoretic approach to, 26
  state-space approach to, 26
  structuralist view, 21
Sex ratios, 108
Sexual reproduction, model of, 115–117
Shepard, Roger, 17–18, 143

Similarity, 21, 50–51, 135, 142–143, 155, 173
  attributes, 145–146
  contrast, 144
  dynamical, 41
  geometric, 41, 143
  kinematic, 41
  mechanisms, 145–146
Simulation, 82
Sober, Elliott, 37, 157–158, 167, 169
Stalnaker, Robert, 137n1
State, 26–27
  initial, 28
  quantum-mechanical, 26
  scope-restricted, 27
  thermodynamic, 26
  total, 26, 79–80, 90
State space, 27–29
  trajectories in, 27–29, 42, 47
Strevens, Michael, 101, 102. *See also* Kairetic account of scientific explanation
Structures, 24, 42, 171
  computational, 29–31
    algorithms, 29
  concrete, 24–25
  construction of, 75. *See also* Target-directed modeling, model development
  deterministic, 44–45
  mathematical, 25–29, 72, 95
  probabilistic, 44–45
Sufficient parameter. *See* Representational capacity
Suppes, Patrick, 26, 95n6

Target systems, 5, 15, 16, 21, 23, 24, 25, 30, 51, 54, 64, 73, 90, 95–96, 114, 171, 172
  abstract. *See* Target systems, generalized
  contingent nonexistence of, 121–124
  generalized, 116–118, 121, 172

Target systems (cont.)
  impossible. See Target systems, nomically necessary nonexistence of
  intersection view of, 116
  nomically necessary nonexistence of, 121, 124–129, 172
  specification of, 88, 172
Target-directed modeling, 5, 74, 112–113
  analysis, 79
    complete, 79–83
    goal-directed, 83–90
  model development, 75–79
  model-target comparison, 90–95
    calibration, 93–95
    goodness of fit, 93, 96
Tasmanian devils, 90, 91
Thomasson, Amie, 52
Thompson-Jones, Martin, 21–22, 48
Three-sex biology, model of, 121, 132–134
Tolkien, J. R. R., 50, 53, 57

Trajectory space. See State space
Transfictional propositions, 54
Turing machine, 130
Tversky, Amos, 143–144. See also Similarity, contrast

Volterra, Vito, 3, 4, 10, 69, 74, 83–84, 150, 159. See also Lotka-Volterra model of predation

Walton, Kendall. See Fictions, Waltonian fictionalism
Watson, James, 24, 38
Weather models, 82, 83, 103, 105
Weighted feature-matching, 5, 147–155, 173
  weighting function, 152
Weisberg, Deena Skolnick, 67–68
Westheimer, Frank, 124
Wimsatt, William, 75n1, 104, 112, 157

xDNA model, 121–124

www.ingramcontent.com/pod-product-compliance
Ingram Content Group UK Ltd.
Pitfield, Milton Keynes, MK11 3LW, UK
UKHW041259180426
11947UKWH00008B/569